THE PHILOSOPHY
OF EVOLUTION

THE PHILOSOPHY OF EVOLUTION

RONALD GOOD

M.A., Sc.D. (Cantab.)

*Professor Emeritus of Botany in the
University of Hull*

DOVECOTE PRESS

First published in Great Britain 1981
by The Dovecote Press
Stanbridge, Wimborne, Dorset

© Ronald Good 1981

ISBN 0 9503518 6 5

Set in Baskerville by
Mouldtype Foundry Ltd., Preston, Lancs.
and printed by Biddles Ltd.,
Guildford, Surrey

Contents

Introduction

Evolution, in its more general meaning of *change with time in continuing directions*, is a fundamental characteristic not only of the planet Earth, but, as far as is known, of the whole cosmos: nothing remains unaltered for long. On earth the mineral part of nature is for ever in a turmoil of transformation and redistribution, while the never-ending repetition of birth and death gives to its living populations constantly changing values.

Most of these processes of change, especially those that affect the cosmos as a whole, are so remote and intangible that it is difficult to appreciate them; and on earth changes in the mineral world are generally so slow that they are not readily perceptible. Only in the living side of nature are they so clearly and peculiarly expressed that they can be closely studied. Because of this the word evolution has, despite its truly ubiquitous application, become particularly associated with the development of life on earth, so that today it is, unless qualified, commonly used to mean change with time in the development of the plant and animal kingdoms. For this reason it is particularly important to remember that evolution in its fullest expression is something much more than this because its omnipresence is reflected in, and gives flavour to, all and every of its consequences; and the history of the living world is so much a direct expression of this as to make clear that however important the evolution of plants and animals may seem to the only living kind capable of appreciating it, man himself, it is indeed but part of something wider and more pervasive.

It is largely to stress this that this book has been called *The Philosophy of Evolution*, a title that carries the implication that it seeks to go rather beyond the scope to which the word evolution is most commonly limited and tries to identify some of the more general principles involved in the comprehensive process of change with time. But because these processes are most clearly apparent and easily studied in their biological contexts, what is commonly called biological evolution, that is to say the story of the development of the world's plants and animals, provides the great bulk of its material.

The evolutionary concept of biology, which rests on a belief that all living things, whether of the past or of the present, are the descendants of prior and simpler kinds and owe their characteristics to a gradual process of slow change over long periods of time, is now so widely accepted as to be almost axiomatic, and an enormous literature has grown up round it. Indeed it may well be felt that there is little left to say, but this is far from true because the more closely the literature is studied the more a great deal of it fails to satisfy the critical reader. There are several reasons for this.

The chief difficulty facing the student of biological evolution today arises from failure during the last hundred years or so to distinguish between evolution in the wider sense and what has come to be called Darwinism. Evolution, of which change with time is the essential expression, is a general principle of nature. Darwinism is an attempt to explain, by means of the Theory of Natural Selection, how the evolution of animals in particular has come about. It cannot therefore be ranked as more than one aspect of the study of evolution as a whole and it is not indispensable for a wider study of the subject.

This conclusion is illustrated in several ways. For example those who seek to pursue an interest in evolution are commonly directed to one or other of the many outline sketches which purport by their titles to deal with biological evolution as a whole but which, in fact, confine themselves predominantly to the evolution of animals in Darwinian terms. The evolution of the plant world, without which the evolution of animals would have been unrecognisably different and perhaps non-existent, is seldom taken seriously into account. This undue influence of the animal point of view in matters evolutionary arises partly from the fact that Darwin, in his earlier years and while he was preparing the *Origin of Species*, was primarily a zoologist, and it is of considerable significance that in his later years he turned much more seriously to the study of plants.

Another point, and perhaps the most important of all, reveals itself in the full title of his book, *The Origin of Species by Means of Natural Selection and the Preservation of Favoured Races in the Struggle for Life*. It is largely from the latter phrase that there have grown up the ideas embodied in two now familiar phrases which, in the last century and a half, have profoundly affected human affairs. These phrases are 'the survival of the fittest' and 'nature red in tooth and claw'*, both of which infer that feral life is a ceaseless battle in which the weakest must go to the wall. Neither of these ideas bears any close examination (see Part II, p. 53 in particular) but they have been made the justification or excuse for many anti-social attitudes and actions and have been misapplied with serious consequences to man.

There are other reasons too why the literature of evolution so often fails to satisfy, and one of the most important of these is that the subject is seldom treated as a discipline in its own right, but nearly always as merely incidental to the study of those other subjects, notably botany, genetics, palaeontology and zoology, to which it is most germane. For this reason the literature tends to be both fragmentary and uneven, great stress being laid on certain aspects of the subject while others, equally important in themselves, are largely neglected. The consequence

---

* This phrase comes from Canto 56 of Tennyson's *In Memoriam*, 1850.

is a serious lack of balance between the plant and animal kingdoms respectively, and there are few, if any, wide-ranging accounts of the subject in which these two sides of nature are treated equally and without bias, and with the proper recognition that whatever the evolution of animals may have been it has been subsequent to, and consequent upon, the evolution of the plant world.

This general lack of balance and correlation is particularly noticeable in the disproportionate amount of attention that has been given to the evolution of the vertebrate animals in general and man in particular. This concentration of interest is easy to understand but it has an unfortunate effect on the study of evolution in general, not only because it is of comparatively little importance in relation to an understanding of the subject as a whole, but because it leads almost unavoidably to an anthropomorphic approach to, and explanation of, many evolutionary problems. This has encouraged an interpretation of biological facts in terms of human beliefs, reactions and emotions that has, in certain directions, gone unacceptably far and which gives quite a false impression of many aspects of the living world.

Another difficulty is that too much attention and argument has become centred on the virtually unsolvable question of what a 'species' may or may not be, another subject which is of little consequence or meaning in the understanding of the realities of evolution. Largely as a result of this preoccupation the study of evolution has long been unduly influenced by what has come to be called 'the species problem'. The effect has been not only to impede fresh thought on many questions but it has made it increasingly difficult to avoid the partisan influences of differences of personal opinion and to see the real problems clearly.

Finally the fundamental question of the origin and early development of life on earth, which lies at the root of all evolutionary enquiries is rarely given the attention it merits. This is partly due to a lack of precise information and partly to the fact that many aspects of the subject lie rather outside the scope of formal biology, but these are reasons for stimulating interest in the matter rather than for neglecting it because until there is some degree of understanding as to how life began and how it developed in its earliest stages so long will our ideas about evolution remain incomplete and potentially misleading.

The Philosophy of Evolution has been written with these difficulties prominently in mind, and it displays its subject by concentrating, first on the facts of nature which have to be accounted for in any theoretical explanation of evolution; then, by surveying briefly some of the more familiar attempts that have been made to explain them; next, by describing in outline the way in which, it is believed, the plant and animal worlds have attained their present state; then, by discussing, as far as is practicable in such a book, the question of the way or ways in which life may have originated on the earth; and, lastly, by summarizing what seem to be the most important conclusions emerging from this treatment of the subject.

The book is the result of a lifetime's interest in evolution, and embodies the thoughts and experiences of many years and in many places. The factual material set out in it is almost wholly derived from sources of general and familiar knowledge and from common observation. This being so it has been felt both unnecessary and invidious to quote many particular sources or references. For like reasons, and because they tend towards that danger of arguing from the particular to the general that so often besets the study of evolution, few detailed examples are quoted, but these can be found in abundance in the sources just mentioned. Much of the book is a logical extension of my *Features of Evolution in the Flowering Plants*, London, 1956 and New York, 1974.

Readers of the following pages will notice that it has often been desirable to distinguish sharply between those plants and animals which have been affected, directly or indirectly, by the actions of man and those that have never felt his influence. To describe the latter, and the state of nature in which they live, the word 'feral' is used in the sense or 'virgin' or 'unadulterated'. Unfortunately the influence of man has now become so widespread that little truly virgin nature remains but the distinction is, nevertheless, a very important one.

One of the most difficult tasks in writing this book has been to assemble the facts and ideas in a consecutive, yet logical, way because the different threads of the argument interweave so much that it is sometimes almost impossible to disentangle them. For this reason also evolution is, outstandingly, a subject which cannot easily be broken down and sorted into convenient pigeon-holes. This is why only a brief and simplified terminal index is provided; but there is an extended contents-heading to each Part.

Finally it must be made clear that the interpretation of the facts described is mine, and mine alone, and I have therefore not included acknowledgments of particular help from others, but it will be evident that the book could not have been written without picking the brains of many people in many walks of life, and to them all I am most grateful.

Particularly am I indebted to Mr. David Burnett of the Dovecote Press for his unstinting and invaluable help in the production of this book.

Albury Park, Guildford, 1981   RONALD GOOD

Part I
The Factual Background

Individuals and assemblages and the diversity between them . . . the axiom that no two individuals are ever quite alike . . . variation between individuals . . . mutations . . . asexual and sexual reproduction . . . flowers and the distribution of sex in them . . . pollination . . . hybrids and hybridism . . . the progressiveness of evolution . . . simplification of form as a response to change in life pattern . . . specialization . . . the absence of true retrogression in evolution . . . orthogenesis . . . selective breeding . . . differentiation between assemblages . . . classification and the necessity for it . . . its limitations . . . the absence of categories in nature . . . the origin of the idea of evolution . . . the time factor.

The evidences of evolution . . . the fundamental unity of life . . . the general parallel between plants and animals . . . recapitulation . . . vestigial organs . . . geographical distribution . . . endemism . . . geographical isolation . . . vicarious assemblages . . . Age and Area . . . differences of dispersal methods in plants and animals . . . conclusions from the evidence of geographical distribution . . . the fossil record . . . the geological time-scale . . . the Great Gap . . . the first macro-fossils . . . the incompleteness of the fossil record . . . the evidence from dinosaurs, ammonites and flowering plants . . . the Cretaceous period . . . further comments on the time factor . . . the historical perspective.

The Factual Background

The plant and animal kingdoms consist (with a few rather debatable exceptions) of an almost incalculable number of individual living things which lead distinct and separate lives and which normally reproduce their kind when mature. These individuals differ from one another in lesser or greater degree and in such a manner that they can be arranged in what may, to use a word free from ambiguity and preconception, conveniently be called *assemblages*; each assemblage consisting of a number of individuals *more like one another than any one of them is like any individual in any other assemblage*. In short there is, among the individuals of both kingdoms, an uneven and discontinuous diversity which makes it possible to recognise different kinds of plants and animals at sight and to describe and name each kind in its own terms.

Assemblages are of all sizes. The number of individuals in some of the commoner assemblages is to be reckoned in many millions; other assemblages are so small that they are scarcely or not at all able to maintain their population from generation to generation, and those that fail to do this gradually become extinct. The total number of different assemblages is very large (most estimates of it exceed a million) and there is a general impression that the total number has been increasing throughout the geological past.

This diversity in nature is most obvious to the human observer in the visible features of plants and animals, but there are also differences, some hardly related to appearance, which cannot be detected so easily. Thus the

9

differences between assemblages and between individuals is both quantitative and qualitative, and although the former can be measured in various ways with some degree of realism most differences of quality cannot, and so the relative values of these characteristics are largely matters of personal opinion.

The recognition of this continuous diversity goes back to the dawn of human history, and many centuries ago philosophers recognised the different assemblages that result, designating the more obvious of them as *species*, using that word in the sense that it has in logic. As an extension of this these 'species' were, as appropriate, grouped together in the higher logical grade of *genus*, though the word 'higher' in this connection does not necessarily confer or imply superiority in any real scale of values. In the context of human language the words species and genus can still be found in common idiom in their original meaning of 'kind' or 'sort' in such expressions as 'mankind', 'something of the sort', 'a version of the story', '*et hoc genus omne*', and most apposite of all 'it takes all sorts to make a world'. However, as applied to the living world these words have, in the last two hundred years or so, acquired senses quite different from those in which they were earlier used, with consequent confusion.

Diversity is most conspicuous *between* assemblages and is of almost every magnitude, but in lesser degree it is to be seen *within* each assemblage, and it is this that gives strong support for the widely accepted belief, amounting to an axiom, that no two individuals, be they plant or animal, are ever exactly alike in all respects. It is possible to imagine, if only in theory, that in those little-differentiated living things which multiply by simple fission into two the progeny of such division might be exactly similar or very nearly so, though even in these cases all experience goes to show that this is not the case, but in multicellular plants and animals the possibility of such a contingency becomes increasingly remote until, in those composed of thousands or millions of cells, the likelihood of total similarity is so small as to be negligible. This applies to any two individuals of an assemblage but it is particularly significant *between parents and their offspring and between brothers and sisters.* As will become more and more apparent in the course of this book the belief that no two living organisms are ever alike in all respects is the ultimate basis of all approaches to the study of evolution and it is essential to understand its full significance before going further. It is a belief which, for obvious reasons, cannot be substantiated by practical demonstration but rests on the experience that no precise facsimiles among living things have ever been recorded. This alone gives the belief a very high degree of probability but the point is clinched by the fact of the passage of time, because this demands that if two individuals are ever to be considered as exact replicas of one another they must have precisely similar life-spans not only in terms of mere duration but, since the passage of time brings continuous change, in terms of simultaneity. It is this factor of time which raises the belief to the status of an axioma medium or universally accepted principle based on experience. The time factor also extends the principle in a direction of peculiar importance in relation to evolution, namely that what applies to the present applies also to the past, so that not only can it be accepted that no two individuals are at any given time exactly similar but also that no individual can exactly resemble any one of its forebears or any one of its descendants. This particular concept of disparity in living nature is the fundamental basis of biological evolution because it means that *every new individual is a novelty.*

The diversity between individuals of any one assemblage is called *variation*, and variation is the key-note of evolution for the simple reason that without it like would beget like generation after generation, the *status quo* would remain unchanged and, in this respect at least, there would be no change with time. It is because parents have progeny which are not just like themselves that the passage of time is accompanied by that change of form which is the essence of evolution.

The connection between variation and discontinuous diversity is clear, and it helps to understand both, as well as the relationship between them, to say that smaller variation may be thought of as discontinuity on a lesser scale, and larger discontinuity as variation on a greater scale. The two are different only in degree; both are comprehended in the dictum that no two individuals are precisely alike; and most of what can be said of the one can also be said of the other.

Variation is most common in the form of very small differences such as are familiar between individual plants or animals of the same assemblage, and to emphasize the dimensions of these differences the word *infinitesimal* has often been used, but less commonly there occurs more drastic variation in which there appear, suddenly and unheralded, individuals so unlike their immediate relatives, especially their parents, that they can hardly or not at all be regarded as belonging to the same assemblage and as meriting the same name. These changelings are called *mutations* (a word meaning change) or in more popular language 'sports'. Both these degrees of variation have played a part in evolutionary theory, but there has been much argument about their relative importance.

This difference of opinion is by no means purely academic but bears directly on the question of the way in which evolution may have operated. One consequence of large differences in diversity between assemblages is that some of the latter appear to be much more isolated structurally than others because there are no intermediate or connecting links between them. Because of this isolation, or specialization as it is more commonly called, such assemblages usually rank, in terms of man-made classification, as 'families'. At the same time many of these families are, within themselves, notably homogeneous, the individuals comprising each showing variation on a single, simple or multiple theme which gives the family its particular features.

This state of affairs can be accounted for in two

different, and largely opposite, ways. One is that it has come about through the long-term accumulation of the effects of small-scale variation, but that in the course of this process many of the antecedents in it have not survived, the result being the major discontinuities now to be seen. The other is that the more markedly distinctive assemblages have arisen in the first place by sudden mutation, the descendants of these subsequently elaborating and proliferating by the slower process of small-scale variation. In short, a mutation showing a novel character or combination of characters appears and becomes the foundation member, as it were, of a new assemblage notably different from any hitherto existing. This progenitor then continues to evolve, chiefly by small-scale variation, producing descendants differing from one another in various details, but all showing the characteristic features of the mutation which gave rise to the new assemblage.

The difference between these two explanations is that the former attaches more importance, at any rate by implication, to variation than to mutation, while in the latter mutation is to be regarded as playing the more important part. This idea has, on occasion, been expressed more vividly by saying that in the former case evolution can be pictured as working *upwards* from less distinctive to more distinctive assemblages, and in the latter as working *downwards* from more distinctive to less distinctive assemblages.

One important distinction needs stressing here. If the first explanation is true then great numbers of different forms representing earlier evolutionary stages (the 'missing links' so often mentioned in the earlier literature) must be assumed to have disappeared without leaving any trace, though there is no direct or satisfactory evidence of this. If the second explanation is correct and the larger discontinuities between assemblages are the results of sudden and considerable mutations, few, if any, intermediate forms would be produced and their absence would therefore not be remarkable. The proportion in which each of these possibilities may have contributed to evolution, and their relative significance as between the plant and animal kingdoms, are still subjects of argument.

Fundamentally variation arises from the fact that living organisms reproduce. These have determinate life-spans, and during them produce progeny which carry on the lineage. The wider significance of this will be commented upon later but for the moment it will suffice to say that, by and large, there are two common methods of reproduction in the living world. One, the less general, is asexual, and consists merely of the division of all or part of a single parent individual into two or more 'propagules' or 'spores'. The other, far more widely seen, is sexual reproduction. In this the nucleus of a cell from one parent individual fuses with the nucleus of a cell from another parent individual and an offspring develops from this double nucleus. Sometimes the two cells from which the fusing nuclei come are alike structurally but more often there is a difference between them, one tending to be relatively large and passive and the other small and active. Where this difference is especially marked the two are called, respectively, eggs and sperms. A survey of the plant and animal kingdoms reveals that this 'differentiation of sex' runs closely parallel with morphological differentiation, and the existence of these two great companion specializations is in itself one of the strongest indications that evolution has occurred.

Some simple forms of life, though reproducing sexually by the fusion of two gametic nuclei, show no apparent sexual differentiation, but in most plants and animals it is possible to distinguish functionally, and generally also visually, sometimes remarkably so, between males and females, or at least between the sex-bearing parts of their bodies. This rather cumbersome statement is necessary because the differentiation and distribution of sex in living things is complicated, but it bears so directly on an understanding of many aspects of evolution that it must be given appropriate notice here. Unfortunately it is difficult to simplify the matter because almost any generalisation about it is subject to exception, but with this proviso the following brief outline is useful.

Broadly speaking the more highly organised the plants and animals concerned the more clear-cut their sexual differentiation is; and in the two great animal groups predominant on land, the arthropods and the vertebrates, the organization into male and female individuals may be regarded as complete. In other animal groups the degree of differentiation varies considerably but, except for some of the simplest, it can be said that separate male and female individuals are commoner than hermaphrodites and, most important, that even in hermaphrodites the sexes are to some degree separated in space or time or in both. Only in some very simple living things are the two kinds of gamete produced in such close proximity that little or no carriage of one to the other is a necessary preliminary to their fusion. The multitude of facts relating to this carriage provides ample evidence for the view that the differentiation and then the segregation of the sexes has been one of the most significant sequences in evolution and that the overall consequence of it has been to lessen the probability of in-breeding and to favour out-breeding.

These same general considerations apply to both plants and animals but the subject is complicated in the former by the prominence of the phenomenon of 'alternation of generations' which, though present in animals, is there much less obtrusive. This difference can reasonably be attributed to the peculiar problems facing passive, immobile organisms in migration from an aqueous to a subaerial environment, a subject which is discussed and described at greater length in Part III of this book.

One of the outstanding achievements of evolution in the development of the land *flora* has been the production of that particular kind of reproductive organ which is called a 'flower', which is the characteristic feature of the third predominant group of living things on land, the

Angiosperms, which on this account are commonly called the Flowering Plants.

It can be claimed with some justification that the 'flower' is one of the most remarkable morphological and functional structures in the whole range of biology and a study of its innumerable and kaleidoscopic manifestations reveals some of the processes involved in evolution particularly clearly, and for this reason it merits careful study. The subject is, however, a complex one and only the briefest outline of it can be given here, but it has been dealt with at considerable length elsewhere*.

Many attempts have been made to describe a 'typical' flower but the enormous range of structure in this organ makes this difficult and all that can usefully be done here is to outline the kind of flower in which the essential organs are most clearly, easily and completely to be seen. Such a flower consists of four different kinds of parts, the sepals, petals, stamens and carpels, each kind separated from the others, and all of them in origin modified foliage leaves. The sepals and petals are purely vegetative in function but the stamens and carpels contain sexual organs producing, respectively, the male and female gametes.

From the point of view of evolution in general and the differentiation of sex in particular the great interest of the flower lies in the fact that the kind just described is by no means the commonest and that most flowers depart from this basic plan in some degree, either in the direction of reduction and separation or in the direction of multiplication and aggregation, or in some combination of the two. At one extreme are flowers comprising sepals and petals and large numbers of both stamens and carpels; at the other are flowers consisting of only one stamen *or* one carpel, this being the absolute minimum that permits of sexual functioning; but whatever the condition may be there is always some spatial separation between the male

---

* Good, R.: *Features of Evolution in the Flowering Plants*, London, 1956 and New York, 1974.

and female organs and this gap has to be bridged in some way if they are to function fully. In bisexual (hermaphrodite) flowers this separation may be very slight but many flowers are unisexual, lacking either stamens or carpels, and here there are two distinct states – the *monoecious*, in which the flowers of each sex are on the same individual plant; and *dioecious*, in which the male and female flowers are on different plants. Uncommonly there are bisexual, male and female flowers separately, either all on the same individual plant or on different individuals. All these three conditions tend to increase the spatial separation of the sexes to a degree depending either on the size of the individual plants concerned or on the actual distance between them, and it is the necessity for crossing these spatial gaps which has produced the enormous body of facts and methods that are comprehended under the title of *pollination*. From the point of view of the student of evolution the most generally important point about pollination is that the methods involve range from the hit or miss procedures of wind pollination to the highly precise use of insect agents in such bisexual flowers as those of orchids and asclepiads.

Normally the two parents concerned in sexual reproduction are so much alike that the differences between them, and the differences between the offspring they produce, amount to no more than small-scale variation, but occasionally the parents are so unlike as apparently to belong to different assemblages. The progeny of such disparate individuals usually show a mixture of features betraying their unusual origin, and are called *hybrids*. It is not easy to say to what extent hybridism occurs under natural conditions wholly unaffected by human influence and most applications of the word are, in fact, to offspring of parents which have become markedly unlike one another through the divergent effects of deliberate selective breeding (see below) in domesticated plants and animals. Cross-breeds of this kind are commonly called mongrels or mules, and are perhaps most striking in dogs, among which selective breeding has been carried to unusual lengths.

Hybridism at first suggests itself as a highly likely agent of evolution, but further consideration leads to a rather different conclusion. First, there is the subjective point that it is difficult to say what deserves the name of hybrid and what does not, because just as there may be great differences in magnitude of variation so there is also a wide range of difference between breeding parents, and, to extend this point to its logical limit, every individual born of two parents can, because no two parents are ever quite alike, be described as a hybrid on a small or minute scale.

A more important point is that the more marked the hybridism the greater is the likelihood, on both morphological and genetical grounds, that the individuals showing it will be sterile, unable to breed with either of their parents or to produce fertile progeny between themselves.

Thirdly there is the point that, in accordance with the Mendelian laws of inheritance, characteristics of small-scale hybrids tend to segregate out in later generations, some of the descendants perpetuating original parental features. This does not in itself preclude the continuation of a hybrid or cross-breed line, and there is the well-known phenomenon of 'hybrid vigour' in which the first few generations of hybrid descent show exceptional vitality, but in these cases the degree of hybridism is usually low, and the hybrid vigour commonly declines to something less extreme in a comparatively short time.

Finally, and perhaps most important, hybridism proper is, in general, uncommon, and the more unlike the parents concerned are, the rarer it is.

On all these counts there is no obvious reason to suppose that hybridism plays any considerable part in long-term evolution, and the impression conveyed by its detailed occurrences suggest that it is, within the general evolutionary scheme, an aberration, and that feral nature is more concerned with its prevention than with its encouragement. There is certainly strong evidence of this in several kinds of plants and animals where the form and arrangement of the reproductive organs in the different assemblages are such as to hinder, if not actually prevent, cross-breeding between them.

That hybridism is not more common invites speculation as to what nature might be like were it otherwise. If, for example, cross-breeding was the rule and not the exception, and parents were more unlike than they commonly are, there would presumably be a situation in which the recognition of assemblages was impossible because of the lack of discontinuity between them. Pursuing this line of thought rather digressively it may be wondered why there is multifariousness at all in the biological world, and why evolution has not proceeded in such a way as to produce but one kind of plant and another of animal, or even just one sort of living thing.

All these three phenomena – small-scale or infinitesimal variation; mutation; and hybridism – have their places in the study of evolution but the first two are reckoned to be the most important. About mutation in particular opinion has fluctuated a good deal. Many pre-Darwinian biologists did not take great account of it, expressing their attitude in the phrase *natura non facit saltus* ('nature does not make leaps') but in more recent times, and especially in the early decades of this century, opinion veered and mutation was held to be of much greater significance. As matters stand today it would be incautious to maintain too dogmatically that either variation or mutation has been the more important process in evolution, though the former is obviously more widespread. Rather they have played different parts of which the relative value cannot as yet be accurately assessed.

A last, but all important, feature of variation is that it gives every appearance of being progressive in the sense that it contributes to change in a continuing direction. This directiveness is, in fact, so evident that it is difficult to realise how different, in theory at least, things might be.

For instance variation might be quite random and without any co-ordination so that it vacillated from generation to generation and resulted in a state of disorder in which it would be difficult to detect any rhyme or reason. Or, variation might be such as to maintain a *status quo* by some combination of action and reaction so that nature would outwardly appear changeless. Or, again, variation might embody a degree of nullification which would produce a negative effect. This last is not only the most difficult to put into words but also the hardest to imagine and there would be little point in pursuing the subject were it not that it has led to some false conclusions which, in turn have contributed to some confusion of thought.

There are many plants and animals that are structurally and functionally much simpler than the other living things with which they are most closely comparable, and this is rightly regarded as having come about chiefly because these organisms lead parasitic, saprophytic or symbiotic lives and have in the course of change with time lost organs and functions for which they no longer have occasion. This opinion is borne out by the occurrence of vestigial organs and bears directly on the doctrine of 'use and disuse'. Since evolution, in the words used above, gives every appearance of being progressive this simplification in certain plants and animals has not uncommonly been taken to indicate evolutionary *retrogression*, and to mean that a reversal of evolution can and does occur, but this view is illogical. A process of progression can only be converted into one of retrogression by a reversal of the progressive trend and a restoration of some *status quo ante*. That there is not, and cannot be, any such reaction is plain enough when it is remembered that evolution is essentially change with *time* and because of this the passage of time alone is a major factor in the process and one which cannot, as far as we are aware, be cancelled out or reversed. Because time passes evolution cannot be retrogressive in the real sense of reversal: there can be no true 'undoing' or 'unravelling' of it.

Unfortunately there has, not infrequently, been a misleading use of the word retrogression, and it has been applied, with the implication of reversal, not only to cases of parasitism and the rest to which reference has already been made, but also to other quite different phenomena. Two examples of this will be sufficient here to illustrate the point at issue. In various insects belonging to groups characterised by the ability to fly the females (and occasionally both sexes) have no wings and this has been ascribed to the results of evolutionary retrogression. Similar views have been put forward to account for the loss of the ability to fly in some birds. The loss of wings may, indeed, be attributed to the effects of change with time but it cannot be the dismantling or demolition of the process by which the wings were acquired because the time factor in it cannot be reversed.

The word retrogressive has even been used in connection with the evolution of the whales with the obvious implication that these animals have returned to a fish-like state of existence by a reversal of the evolutionary process that, earlier, led to the development of land mammals from aquatic vertebrates, an idea which sometimes receives reinforcement from the occurrence, in the whales, of certain vestigial limb-bones.

Misunderstanding here has been caused by a failure to distinguish between evolution itself and the specialization which is commonly its concomitant. There seems no reason to quarrel with the opinion that the wingless animals referred to have, in the course of time, lost their wings because they have gradually adopted a life-style, or occupied a habitat, in which wings are no longer of their original value to them – they are able 'to get along without them' – but these animals are *more*, not *less* evolved than their congeners. The whales make the point even more plainly: they are mammals and perhaps the most striking of all examples of specialization in a group singularly heterogeneous in both form and mode of life. They are mammals which have exploited their own particular kind of habitat and this has been possible because evolution has provided them with the special form and function most appropriate to this exploitation, and in the process they have undergone an accumulation of change greater than that common to other members of their group. To suggest that specialization of this kind is an unwinding of change with time, and thereby a reversal of evolution, is not warranted.

Biological evolution is an irreversible process in the sense that no inter-generational change, such as is the basis for the conclusion that every new individual is a novelty, can ever exactly reverse one that has previously occurred. Inter-generational change may have the effect of diminishing the value of this or that particular characteristic but it cannot so completely reverse the values of these as to restore any previously existing state. This is partly because to do so would involve simultaneous and appropriate changes in many, if not all, the characteristics concerned and partly because time is all the while elapsing. Time so passed cannot be recaptured or annulled, and therefore no two situations or conditions separated in time can ever be truly equivalent. The belief that biological evolution is irreversible has therefore the same claim to the status of an axioma medium as has the belief that no two individuals are ever exactly alike, and it is similarly based, not only on theoretical argument, but on general experience.

The tendency for variation to persist along particular lines, and consequently to produce specialization of form and function, has long been recognised by biologists and has given rise to the concept of *orthogenesis*. By this word is meant continuing evolutionary change in certain directions determined by inherent factors rather than in more numerous and divergent directions controlled by external factors. Only by its most fervent exponents is it suggested that orthogenesis is the master key to evolution, and it cannot be regarded as a prime cause of it, but hints or suggestions of it emerge over and over again in evolutionary studies, and indeed progressive variation presupposes it. It has so far proved difficult to express the concept of orthogenesis in the form of a concise theory but, as the longer reference to it in Part II shows, it is something that must always be borne in mind as an important element in evolutionary knowledge.

The idea behind orthogenesis is well illustrated by what is called *selective breeding*. Long before the nineteenth century it was well known among those concerned with the breeding of domestic plants and animals that it is possible, given enough time in the form of a sufficiency of generations, to improve almost any breed by selecting, as parents of the next generation, individuals showing most strongly in themselves the characters which it is desired to emphasize in their offspring, because experience shows that these characters thus become more and more accentuated from generation to generation. This selective breeding shows clearly that, in certain circumstances, in this case the agency of human choice, variation does accumulate orthogenetically. This important conclusion leads to a crucial issue regarding evolution. Is there, in feral nature, some process akin to selective breeding but quite divorced from the actions, and even the existence, of man?

Just as no two individuals are alike so also no two assemblages can be regarded as alike in all respects and as exactly equivalent to one another. This differentiation *between* assemblages is most clearly revealed in their structural features and, as will be clear later, it is this which makes it possible to classify them, but they also differ in one or more of many other ways. For instance no two assemblages have exactly the same geographical distribution, if for no other reason than that no two individuals can occupy exactly the same space. Much the same is true of the number of individuals each assemblage contains, but here the criterion is one of finite numbers, and the possibility that these may, especially among smaller assemblages, be the same cannot be ruled out, but even if this should be the case at any one time it would be temporary only because the number of individuals in any assemblage tends to vary from generation to generation. Assemblages also vary in their reproductive capacities as well as in many other directions, but there is one respect in which the situation is not quite the same for both assemblages and individuals.

This is in their relationship to their environments. It is a commonplace of nature that assemblages, though they may be superficially similar, often occupy different kinds of habitat and are therefore inevitably geographically segregated, if only on a small scale, and on this account the word 'vicarious' is applied to them. This situation finds no close parallel among individuals because individuals which occupy any particular kind of habitat usually reveal this association in their external form and in consequence are recognised as distinct assemblages.

The foregoing brief survey of the principal facts concerning the organization of living nature which must be satisfactorily accounted for in any acceptable explanation of evolution and the way it works, has an engaging simplicity which raises the hope that further enquiry into them may be equally straightforward. Unfortunately this is not the case, and it is important to understand why this is so. It is because there is involved a limiting factor which, as the coming pages will repeatedly demonstrate, makes itself felt in almost all discussions about evolution. This factor is that the powers of human understanding, communication and expression are often inadequate and need to be supported artificially. The vast numbers and diversity of plants and animals present just such a situation and the human mind can hope to comprehend them fully only by devising and using some simpler representation of them by which they can more conveniently be described, dissected and handled.

This necessary simplification is achieved by applying the process of *classification*. The essence of classification is no more than the combining together of many small entities into fewer larger ones or the dividing of larger unities into more numerous smaller ones. It is applicable to almost every aspect of man's mental and physical activities and has no particular association with biology, but circumstances have combined to make it of special significance in this field of study, and today a great deal of biological thought is expressed and measured in the language of classification.

Plants and animals have been classified in the simplest terms since the earliest historic times, but more modern and elaborate arrangements are particularly associated with the Swedish biologist Linnaeus (Carl von Linné) who lived from 1707 to 1778 and whose most important work was done between 1730 and 1750. His systems of classification are regarded as the foundation of modern *taxonomy*, as the study of classification is now called. This word should be carefully noticed because it has two useful connotations. It comes from the Greek word 'taxis' which, it is interesting to learn in view of the example given below, was the name of a military unit, and it has given rise to the useful word 'taxon' by which any single entity of a classification, irrespective of the category in which it is usually placed, can be described.

The classification of a very great number of different facts must, if it is to fulfil its purpose adequately, almost inevitably be hierarchical, by which is meant that it must consist, on the general principle of 'the higher the fewer', of a series of ranks or categories each theoretically superior to, and comprising members of, that below it. Hierarchical classifications are most familiar in the sphere of human society and most simply illustrated there by chains of command and responsibility, and their accompanying prestige, such as exist particularly in the armed services. Here the entities to be classified are individual human beings. As with all other individuals no two of them are, in fact, alike but for purposes of organisation they are treated as equivalent to one another and susceptible to hierarchical arrangement. At the lowest level is the 'rank and file' and above them is a series of superior commanders so that the whole arrangement comprises a number of levels or ranks, each more prestigious but containing fewer members than, the one below it. In this way the ubiquitous diversity between the individuals is masked and translated into a simple framework of words which, for practical purposes, can be used as a sort of shorthand. In biology classification operates in the same manner except that the commonly used categories or ranks are fewer, generally only four – the *species*, the *genus*, the *family* and the *order*.

Classifications of the kind just described provide a working tool and a vocabulary without which biologists would be sadly handicapped and they are usefully employed even by those most aware of their unreality, but it should be recognised that they have two inherent limitations.

The first, and fundamentally more important, is that biological classification is based upon and perpetuates the fiction that assemblages can be arranged in real categories, the members of each of which are equivalent to one another. That this is not so is clear from the very fact that no two assemblages, or of the individuals they contain, are alike, and the assumption that they are so gives a fundamentally false picture of the plant and animal worlds.

The second limitation is rather different in kind and by no means confined to scientific biology. It is that while biological relationships, like other genealogies, are at least three dimensional, they can only be described in speech or writing in two-dimensional terms, that is to say one after the other, and within the limits of the written or printed page. This being so sequence is the essence of them, and given this it is almost impossible to avoid the idea of precedence, because, to describe the facts, it is necessary to make lists of them, and lists can only be made by putting items in succession. Attempts have been made to lessen this disadvantage by expressing the facts in three dimensions but this is complicated and has not been widely adopted. Classifications are therefore in most ways lists in which the items are dealt with one after another and this leads almost inevitably to the ideas of 'priority' and its concomitant 'superiority'. In more practical terms it offers an almost limitless opportunity for rearranging the order in which the items appear according to personal opinion.

It is reasonable to suppose that when Charles Darwin used the word 'species' in the title of his great book, *The Origin of Species* (see Part II) he did so in its original meaning (see p. 00), but when, largely as a result of his own writings, the idea of evolution became more acceptable, the word began to lose its pristine meaning and came to be used, not merely to describe some recognisably distinct collection of individuals, but in the sense of a category in a hierarchical classification, and therefore as indicating a certain status, rank or stage in the process of evolution. In this way the word 'species' became, as it were, a standard measure of classification having a relative value of its own, and this it has, in practice, ever since remained.

Unfortunately no reliable means of quantifying this measure has been found, nor, in purely biological terms, is it ever likely to be, and so it is impossible to say what, and what does not, constitute a taxonomic species. Not only this but in the last hundred years or so there has gradually grown up an opinion, if not a belief, that the kind and degree of morphological difference which permits individuals and assemblages to be classified into groups are, in themselves, some indication of the degree of genealogical relationship between them – that pedigrees can be discerned and traced from purely formal resemblances. This idea can be a valuable help from some points of view but it affords a perilous sea on which to embark because the values of likenesses and differences in the living world can only be calculated (as distinct from being described in words) by applying to them human estimates of relative importance, and since these are bound to reflect purely personal opinions they must not only be regarded with great caution, but they easily become the subject of sterile argument.

Not only this but they lead almost unavoidably to notions of relative worth and value, and these import into the study of biology the wholly misleading and inapplicable ideas of 'better' and 'worse'. This distorts the whole true picture of the living world because, difficult as it may be for man, accustomed to evaluation and hierarchical classification, to understand, *nature knows no categories.* Every individual or collection of individuals is sufficient unto itself and has its own particular combination of qualities, and although any of these may be more or less alike there is not, and in nature cannot be, any element of status or precedence in comparing them, other than such as may come from subjective assessments made by the human mind. The words 'better' and 'worse' have real meaning only in relation to the achievement of some end or purpose, and this in turn involves the notion of 'desiribility', and since no one knows what, if anything, such a goal and its nature and its influences on the living world may be, the ideas behind these words are wholly inappropriate to feral life. Hence, hierarchical classifications, and above all the concept of the 'species', though they may be essential as working tools, must not be regarded as more than this. In particular, although they may help to point the way to some of the evidences of evolution they should not be regarded, in themselves, as affording evidence of it.

It is so important in the study of biology to realise the absence of categories in nature that this demands one further brief comment. At the beginning of Part II and at the end of Part IV of this book it is suggested that the first steps in the development of that intelligence by which man is distinguished from all other living things were the acquisition of the powers to associate cause and effect and to be conscious of time past, present, and future. To these may be added, if indeed it is not already inherent in them, a third acquisition, namely the power of estimation in terms of relative 'value', such as is expressed in the words 'better' and 'worse'. Once this power is acquired there surely must follow the opinion that, in some respects at least, usually typified as benefit or detriment to man, the individuals of an assemblage are unlike one another not only in infinitesimal variation or mutation but also in value, whatever that may mean, among their fellows. This can very soon become assimilated into acceptance of the opinion that some individuals and assemblages are more 'worthy' than others, and from this it is but a short step to the idea of 'class'; and the mis-handling of this wholly man-made concept has provided much of the less praise-worthy material of human history. In truth feral living nature provides the prototype of a community totally classless in this kind of sense.

But this account of classification, important as it is, has been something of a digression and we must now return to the main theme of evolution and particularly to one of the fundamental questions about it, which is how the idea first arose. For many centuries it was believed that plants and animals had come into being through or by one or more acts of deliberate creation on the part of some super-natural agency, variously identified and described; but in more recent times, partly because of the immense array of new facts which had to be accommodated within any explanation, and partly because of the general change in attitude of mind away from the mystical towards the more realistic, opinion came more and more widely to hold that living nature as it is seen today is the result of natural rather than super-natural causes, and this growing opinion has, over the years, crystallized into a belief that it is the result of a process of slow unfolding and gradually increasing elaboration from simple beginnings – that it is due to progressive evolution.

It is necessary to use some such word as progressive here in order to distinguish the supposed process from many others to which the word evolution can, in its broadest sense of 'change with time', be applied. For example the erosion of rocks involves the separation, transportation and reassembling of material from one form or place to another, but this is essentially a rearrangement or existing matter and has no element of progression such as is embodied in biological evolution. Similarly the gradual elaboration of human inventions can properly be described as evolutionary, and here indeed there is an ingredient of progress, but of a kind unlike that of biological evolution because the subjects are not animate and do not reproduce themselves.

The great difficulty in studying biological evolution is that the process has involved vast periods of time and so, to our eyes and minds, appears to be an exceedingly slow affair, so slow indeed that its major operation cannot be directly observed. In consequence the evidence for it is circumstantial and a belief in it rests on a conclusion that this evidence is sufficient. Because of this it is often said that the idea of evolution cannot be more than a theory, and this is strictly true, but it is too austere an attitude from which to work, and at least for all practical purposes of study it is justifiable to regard the evidences of evolution as strong enough to merit the belief that it is, indeed, a general principle of the living world and the explanation of its multifariousness, though not necessarily of its origins. What these evidences are must now be reviewed, but as a preliminary it is worth while to recall the conclusion just expressed that evolution is a general feature of the natural world, because, if this is true, there is a *prima facie* reason for believing that the living part of nature is no exception. Observing, as we may, the ubiquity of evolution in one expression or another it would be more remarkable if it did not apply to the living world than if it did.

What are usually called the 'evidences of evolution' are many and can best be set out under two headings, one comprising those evidences deriving from knowledge of the living things inhabiting the world today; the other from what is known about the plants and animals of the past.

As regards the first the most important evidence is that afforded by the fundamental unity of life wherever it is to be found, as is expressed most plainly in the presence of protoplasm; in the cellular structure of living bodies; in the ubiquitous occurrence of variation and discontinuity; in the general methods of reproduction; and in many physiological processes, such as are particularly illustrated by respiration. The last of these, though the least obvious to observers, may be considered the most significant.

This overall similarity suggests strongly that all living things have come from one and the same original source, and that they are therefore all members of one and the same great lineage. If this is so then the likelihood that the present state of affairs has been brought about by an evolutionary sequence is very strong.

Besides this general similarity there are also various particular parallels between members of the plant kingdom and members of the animal kingdom. Not all of these are easy to describe in words and, for their appreciation a wide knowledge is necessary, but it can be said without much fear of contradicton that biologists reasonably well acquainted with both plants and animals cannot fail to be impressed by the general correspondence between the main groups of the one and those of the other. To cite but one example that is especially apparent in land plants and animals, a leading developmental theme in both kingdoms is the greater preparation and protection of their progeny.

The more general evidences of evolution are reinforced by certain secondary and less ubiquitous facts of which two especially have been the subject of much discussion, chiefly but not exclusively among zoologists. In the predominantly land-living groups of the vertebrate animals (amphibia, reptiles, birds and mammals) the young, during their early development commonly pass through stages which are reminiscent, if no more, of the adult condition of groups generally regarded as 'more primitive'. This phenomenon is called *recapitulation* because it seems to indicate that the embryonic or pre-natal stages of these animals correspond to, and reiterate, the adult conditions of the ancestral forms from which they have descended. This idea has been summarised in the epigram or dictum that 'ontogeny' (which is the development of the individual) repeats 'phylogeny' (which is the history of its kind). Many of the facts relating to this idea are striking and the inference from them may well be sound, but as evidence of evolution in general it would be stronger if its most obvious expression were not so particularly confined to certain highly developed land animals.

Much the same can be said of another morphological feature which is generally regarded as evidence of an evolutionary past. This is the presence, in various animals, of *vestigial organs*, the appendix in man being perhaps the most often quoted. It is supposed that these are the reduced or atrophied successors of organs which, in the ancestral forms of their present possessors, were fully functional but which, because of changes in mode of life and consequent structural modification, have lost their former usefulness and have gradually 'withered away'. Here, again, the inference is probably sound but the phenomenon is not sufficiently widespread to be more than an ancillary and supporting suggestion of evolution. Something of the sort can, by a stretching of the concept, be detected in plants, but it is important to notice an essential difference here. The metabolism of plants is, compared with that of animals, so basic and unvaried that there is far less scope for the kind of nutritional or ecological change that is held to be the explanation of vestigial organs in animals, a point which particularly reflects the 'activity' of animals in contrast to the 'passivity' of plants.

Darwin was fascinated by the geographical distribution of plants and animals, and ever since the publication of *The Origin of Species*, in which considerable space is given to it, the geography of living things has provided a lively debating point in the study of evolution. The facts about it are multitudinous and varied but the literature is copious and easily available and a few general remarks are all that is needed here.

First and foremost it is to be remembered that the present distribution of living things is not, in itself, necessarily evidence of an evolutionary population of the world; the same result could have been achieved by some process of 'special creation' (see Part II). Nevertheless when the detailed facts are considered it is soon realised that to produce them special creation would have had to be of a complexity difficult to imagine.

The aspects of distribution that lead most particularly to this conclusion are, broadly, four. The first is *endemism*, the occurrence of assemblages confined to one particular, and usually relatively small, tract of country. The second, closely related to the first, is *geographical isolation*, the restriction of assemblages to geographically or environmentally isolated areas such as oceanic islands or isolated mountains. The third is the occurrence of closely similar *vicarious assemblages*, or geographical races as they are often called, in different parts of one and the same, usually latitudinal, climatic zone.

The fourth aspect, which is in fact the most widespread, cannot be expressed quite so concisely, and needs a longer word of explanation. It is based on the observation and experience that, just as no two assemblages are exactly alike in form, no two have exactly the same geographical distribution. At the same time it is well known that the constituent assemblages of larger groups, or the genera of a well-defined and homogeneous family, show considerable differences in the extents of their distributions. This state of affairs has, because plants are essentially passive and without the power of locomotion, been especially studied by botanists, and from these studies there has emerged the hypothesis to which the phrase *Age and Area* has been applied. Certain aspects of this and of the conclusions that might be drawn from it have generally been controversial, but from the point of view of evolution it is of great significance, because, with certain reservations it suggests that, at least among closely similar forms with comparable powers of dissemination, the extent of distribution is a measure of the relative age of the units concerned, the older having had more time to spread and having therefore the wider ranges. The important point here is the time factor because, if the general idea of Age and Area is sound, it is strong evidence that, of the assemblages involved, some have been in existence longer than others, and this can most easily be rationalized in evolutionary terms.

Another comment on geographical distribution is rather different and draws attention to the fundamental distinction between the methods of dissemination in plants and those in animals. That plants are to be found in every part of the world where neither frost nor drought inhibits their active growth is proof enough that their means of dispersal are fully adequate, and the same is true of animals, but their case is basically different because of the powers of mobility, volitional or other, that most of them possess. It is therefore important not to equate too closely the geographical distributions of plants and those of animals. Nevertheless those of animals, because they must preserve congruity with the environmental conditions needed, must follow those of autotrophic plants. It is the difference in method between the two and the fact that the spread of plants precedes that of animals that must be borne in mind in any comparison between the two.

The facts of plant and animal geography seem to show, beyond reasonable doubt, that whether special creation was or was not the primary cause of the population of the world, there has been, since that peopling first occurred, the kind of change with time that is called evolution. This point is of particular interest because it puts into words the possibility of a combination of special creation *and* evolution. Indeed, if special creation was the original cause then the subsequent changes in the mundane environment which we know, or assume, to have occurred, would in the highest probability have been accompanied by changes in the forms and functions of living things so created, and this would be evolution.

The second kind of evidence of evolution, that provided by history, is the more easily appreciated and here the circumstantial evidence is strongest. This is the indication afforded by what is called the fossil record, that throughout much of geological time there has been a continuing succession of different kinds of plants and animals. The sedimentary rocks which now form much of the surface of the earth have been laid down over a vast length of time, and geologists, by comparing adjacent or neighbouring deposits in different lands, have arranged them in a sequence which, at least in broad outline, expresses the order of their times of formation, and a comparison of the fossils in these sedimentary rocks reveals that, generally speaking, the older the sediment the simpler and less highly differentiated the plants and animals whose remains are entombed in them.

The whole gamut of the geological time-scale is usually divided into five parts of unequal duration. First, and incomparably longest, is the Azoic series of rocks, so-called because they contain no fossils and are therefore thought, *either* to date from before the coming of life *or* to have become so changed or 'metamorphosed' since that any evidence of life in them has been destroyed. This is the least definable of the five parts because unless it is assumed that life has existed on earth as long as the whole geological sequence there must have been an original period during which rocks were being formed or laid down but before any living things appeared, and what the duration of this pre-organic stage may have been we have little or no idea, except that it must have been immensely long and, in all probability, longer than the whole subsequent organic stage, but there is an immense potential latitude in any attempt to estimate the time involved.

Next comes the Proterozoic or Archaeozoic series, in which there are some fossil remains but only of comparatively simple marine forms of life. These are followed in succession by the Palaeozoic; Mesozoic; and Caenozoic eras, and it is these three that constitute the fossil record proper. The Palaeozoic is divided into six parts and during it, probably nearer the end than the beginning, the first fossils of land plants and animals appear. Expressed in botanical terms the Palaeozoic contains fossil floras in which plants of the fern kind (Pteridophytes and Pteridosperms) are the highest expression; the Mesozoic, which is divided into three, reveals floras in which the earliest seed-plants (Gymnosperms) are the most characteristic element, though flowering plants made their appearance before its end; and the Caenozoic, which is also subdivided, is the epoch of the Flowering Plants in overwhelming preponderance.

That successive rock deposits contain fossils of ever-increasing complexity of form, and that plants and animals of the land do not appear until comparatively late in the sequence, would seem to provide almost conclusive evidence of biological evolution, and it is widely accepted as such, but there are certain points about it that need to be properly understood.

The most obvious of these, though often the least realised, is that the fossil record is not *proof* of evolution because it might equally have resulted from special creation. It provides what is usually regarded as strong evidence for it; but an even stricter statement is that the record is highly suggestive of evolution. This was roughly the attitude prevailing in the early years when the fossil record was gradually becoming plainer and evolution was beginning to be regarded as the most likely explanation of it.

The second point, which bears directly on the first, is that an immense length of time must have elapsed between the appearance of the first living things and the date of the first macro-fossils. Apart from some problematical remains the earliest fossils familiarly known are the graptolites and trilobites of the earlier palaeozoic rocks. The former are members of the Hydrozoa, belonging to the Coelenterata, and the latter are arthropods, and why these two very different types of animal should be so well known from the rocks of their age is difficult to understand, and it must be assumed that circumstances, at present unknown, prevented the preservation of other contemporary types, for these two cannot be held to represent the only animals of their time. The trilobites are relatively highly organised creatures comparable with some of the Crustacea of today, and that this level of development should have been reached so early in what we know as the fossil record suggests that the length of biological time before it was even greater than is commonly supposed on other grounds. It also suggests that the time which has elapsed *since* these first fossils were entombed, embodying as it does the whole colonization of the land, has been much shorter than the unknown ages before it, about which we have virtually no information and which, for this reason, may well be described as the *Great Gap*. Because of it very little is known about two crucial stages in evolution – how animals became differentiated from plants, and how bottom-living plants and animals (the former attached and the latter for the most part mobile) came into existence. More generally it obliges us to interpret evolution too exclusively in terms of what has happened since the beginning of the fossil record as it is known today, and how far it is sound to do this it is not possible to say.

The third point concerns the completeness or otherwise of the fossil record. Does the succession of fossil-bearing rocks represent the whole passage of geological time since the earliest of them or have there been times when no contemporary plant or animal remains have been pre-served, or deposits formed which, though once fossiliferous, have been destroyed by subsequent action? Certainly the disconformities between certain adjacent geological for-mations, and especially between the fossils they contain, are conspicuous, and indeed it is the very existence of these that allows the record of the rocks as a whole to be divided into chapters in the way described above, just as it is the morphological discontinuities between assemblages that permits biological classification. It is also generally thought that some breaks in the record are probable because of the absence of certain 'missing links' or intermediate forms. Nevertheless it has long been rather tacitly assumed that such gaps, if they exist, are comparatively small and local. Interest in them however has quickened in recent years and the idea of gaps in the record has been reinforced by hypotheses of varying merit and practicality.

Attempts to explain what appear to be gaps in the record tend to concentrate on the particular case of the Dinosaurs, the great reptilian group which dominated the land fauna during the Mesozoic and which are usually described as disappearing at the end of that era, though some accounts feature them as continuing into the early Tertiary. Chiefly because of the very large size of many of them, and the consequent dimensions of their fossil remains, these animals have attracted much attention and it is well to remember that some other animal groups, notably the ammonites, are comparable in this respect. It is, however, the great plant group of the Angiosperms, now dominant in the vegetation of the world, which illustrate the problem most precisely. There are two great groups of these plants, the Monocotyledons and the Dicotyledons, the former being of special note because they include the grasses and the palms. The Dicotyledons are credited with a mid-Tertiary origin, but the Monocotyledons appear to be older and some of them are thought to have existed contemporaneously with some of the Dinosaurs. This is important because, on various structural grounds, it has been suggested that the palms may have had an origin different from that of the rest of the Angiosperms and may not be wholly homologous with them. It is, however, the Dicotyledons, which contain nearly all existing woody land plants, that most amply fill in the picture. How and when the Dicotyledons came into being is still very much of a mystery and all that is known with some certainty is that their fossil remains are first clearly recognisable in rocks classified as of Cretaceous age. Most of these fossils are leaf impressions but there are also some fruits and other structures and all go to show that the plants concerned must have been scarcely distinguishable from some of those now living. This is perhaps the most important point of all in relation to evolution in general because it seems to show that, since their earliest times there has been comparatively little change in many kinds of Dicotyledon, though no doubt the numbers of their assemblages has increased, and this in turn indicates that the time which has elapsed since the Cretaceous may be inconsiderable compared with that which saw the origin and gradual proliferation of this group of plants.

In short, taking all the plant and animal groups mentioned into account it is difficult to resist the conclusion that there must have been a considerable hiatus in the geological record towards the end of what is commonly called the Cretaceous. Similar lines of evidence, though less picturesque can be detected in other biological groups and geological periods and it seems plain that much of the evolution of the land biota must have taken place during these gaps and that the fossil record is, to that degree, incomplete.

These comments on the fossil record lead the enquirer very directly to what is at once the most important and most indescribable element in biological evolution, the ingredient of time, to which already in these pages many references have been made. A book on biological evolution is not the place in which to go deeply into the metaphysics of the subject but it does give an opportunity for emphasizing some points. First and foremost all that can reasonably be said about time in relation to evolution is that the latter proceeds at a rate which makes it impossible for man to observe its operation directly. So, as far as he and his traditions are concerned living nature appears to be more or less unchanging, and it is this abiding impression that led early man towards the original idea of creation and, later, to its proliferation into various 'creation myths'. It is true that there are, on occasion, the sudden evolutionary changes that have come to be called mutations but how far these control or contribute to the grand march of biological evolution is by no means clear.

All this being so it is a natural conclusion that the duration of time, both past and present, is something of immense dimension and, since man yearns for figures and statistics because they appear to give body to what otherwise seems incorporeal, there have been persistent attempts to express the dimension in figures. Inevitably these estimates are couched in terms of the largest numerical unit in common use, the million, and some idea of time has long been expressed in millions of years. Used with reason this is probably the best that can be done but recently the use of year/millions has been overdone and it is hard to find any outline account of evolution today in which there is not a continuing accompaniment of hypothetical age figures. In consequence the sceptic tends towards the impression that these are too often used as cloaks of ignorance by which lack of knowledge is concealed. In any case it is salutary to remember that to talk of millions is to use a language which has a considerable degree of incomprehensibility.

It is however helpful to bear in mind what some of the figures imply in terms of generations. To return to the case of the Angiosperms these plants are generally credited as having become dominant at the end of the Cretaceous or the beginning of the Eocene, a stage in the earth's history commonly dated at about 70 million years ago. Most Angiosperms, and especially Dicotyledons, reproduce, when mature, at least once a year and if this estimate is sound there must have been at least 70 million generations during which the features of at least some and probably many of these plants have scarcely changed. This is perhaps as close as one can hope to get to an appreciation of the speed of evolutionary change.

These comments on time past lead directly to the question of the rate or speed of evolution, because the fossil record affords cogent evidence that this has not been constant in terms either of time or classification. Beyond this bald statement it is difficult to go because much evolution involves changes which are neither directly nor indirectly perceptible to the eye and which therefore cannot be measured in any normal way, and little more can be done here than to make one or two generalizations.

Since all new individuals are novelties unlike any of their forbears or contemporaries the speed of evolution will depend largely on a combination of two factors, the *degree* to which offspring differ from their parents and the frequency of reproduction. If the differences are small and such as to produce little change in external form an original type may seem to continue almost unchanged for long periods of time. Some of the Brachiopods, which are known from Archaeozoic times, are most commonly quoted as extreme examples of this, though it would be more precise to say that they *appear* to have undergone little or no *external* change in that time. If the intergenerational changes are large, perhaps approaching the dimensions of mutations, the rate of evolution will be appropriately faster. As to the frequency of reproduction it may be doubted whether, over the whole biological spectrum, there is sufficient range of difference in this respect to make it an important factor because the longest periods between successive reproductive periods are negligible in terms of cosmic time. Nevertheless it is important to distinguish between intergenerational variation which may, in the longer-maturing plants and animals be a matter of many years, and the much more frequent repetition of reproduction by the same parents which, in broad terms, is seasonal.

One aspect of the fossil record that must certainly be noticed here is not particularly geological. To the human mind the past is like a long range landscape in which the nearer the object is to the observer the more clearly it appears and the greater its importance to the whole vista seems. All human history provides examples of this but formal geology illustrates it especially well because the more recent the past the more that is known about it. Thus, in the Caenozoic era, which is the age immediately preceding the present, there is first a division into two,

Tertiary and Quaternary, and then a further division of each, the former into four and the latter into two, but these last two represent only a minute fraction of the total time that is thought to have elapsed since the beginning of the Palaeozoic. This mental telescoping of the past has a distorting effect on our appreciation of the facts and due allowance must always be made for it. How much our opinions would change if we knew as much about the more remote past as we do about the immediate past, and could dissect it in the same detail, is at least an interesting speculation.

Finally, consideration of the fossil record leads directly to the question of the origin of life, which is the subject of much of the fourth part of this book. Is the record that of a single biological lineage or not? If it is, then the innumerable plants and animals of the past now known only as fossils are the blood ancestors of those now living, and a belief that the record is a demonstration of evolution is immensely strengthened. Unfortunately this has never been, and perhaps can never be, decided, and so it cannot be accepted as an established truth. In consequence the fossil record, however complete it may appear to have been, is, strictly speaking, not conclusive evidence of evolution.

Part II
Theories of Evolution

Man's gradually awakening awareness . . . 'creation myths' and 'special creation' . . . Charles Darwin and The Origin of Species *. . . the Theory of Natural Selection . . . the factual basis of the theory . . . the general stability of the populations of assemblages . . . 'good' and 'bad' seasons . . . pests and plagues . . . the environmental control of populations . . . the over-production of propagules . . . the equilibrium of feral nature . . . the analogy of cultivation . . . the false idea of 'better' and 'worse' in nature . . . the difficulty of applying the concept of natural selection to plants . . . difficulties arising from the use of certain words in Darwinian literature . . . sexual dimorphism and the Theory of Sexual Selection . . . Lamarckism and the Theory of Use and Disuse . . . the 'inheritance of acquired characters' . . . superficial resemblance . . . 'protective resemblance' and camouflage . . . the sense of sight in animals . . . melanism . . . 'mimicry' . . . 'warning colouration' . . . sensory perception in animals . . . the concept of 'merkwelt' . . . evolution pictured as a much-branched tree always almost totally submerged by the rising tide of time . . . the analogy between orthogenesis and a railway system.*

Theories of Evolution

Ever since man became conscious of his environment and of its astonishing diversity he has been able, because of the essential discontinuity of nature, to recognise the different kinds of plants and animals and, at least in the case of those most important to him, to give them names. As his intellectual powers grew so his consciousness of them increased and he began to wonder how the state of affairs that he could see around him had come about. No doubt this wakening awareness was a very slow process and no doubt also it was from the beginning bound up with the development of theism, because his readiest explanation of facts which he could neither understand nor explain otherwise was that they had been produced through the agency of some super-natural power, the nature of which he could imagine but faintly. Later on, as simple theism crystallized into various formal religions, this picture was amplified in different ways until most of these religions came to have, as integral parts of their dogmas, a story intended to explain, in simple words, the origin and appearance of life on the earth. These 'creation myths', as they have been called, are of many kinds, some more bizarre than others, but the basis of many of them is the idea that each sort of plant and animal came into being by an act of creation; that they were 'specially created' in this way. Exactly what the relationship may be between the word 'species' as used by the ancient philosophers and the word 'specially' used in creation myths is not entirely self-evident, but by the mid-nineteenth century, at least in the Christian world, they bore the relation of noun and

45

adverb, and at that time a belief in Special Creation was the orthodox and generally held explanation of the origin of life on earth.

This is not to say that there were none who thought otherwise, and many biologists as well as other thinkers had, in the face of the many facts which continued to come to their notice, grave doubts about the validity of a belief in this explanation, and were groping their way towards some other, and almost inevitably evolutionary, reason for them, but it was not until 1859 that their doubts began to find effective expression through the publication in that year of Charles Darwin's book *The Origin of Species by means of Natural Selection and the Preservation of Favoured Races in the Struggle for Life.*

Darwin's purpose in writing the *Origin* (as the book is commonly called for short) was twofold. He sought to demonstrate by argument and example that species are not immutable but are produced by a process of gradual change from generation to generation and, by inference from this, that the different assemblages of plants and animals have not been 'specially created' but have evolved from and through their ancestors. The principal agency of this proceeding he described as 'natural selection'. The effect was to relate the general concept of evolution and the more particular idea of 'natural selection' so closely that they have ever since been intimately, and sometimes inextricably, interwoven. It is therefore of prime importance to consider how far this close association is realistic and justified.

The Theory of Natural Selection, which was propounded independently and almost simultaneously by Charles Robert Darwin and Alfred Russel Wallace, postulates that the plant and animal kingdoms have evolved from ever-simpler ancestral forms by a process analogous to selective breeding by man but actuated by natural forces operating over long periods of time. Since no two individual living things are ever wholly alike some will always be more in tune with the conditions under which circumstances oblige them to live than others, and these will survive to reproduce their kind while others, less appropriate to these conditions, will perish. This process repeated generation after generation will eventually result in individuals so different from their progenitors that they can reasonably be held to constitute a new 'kind' or 'species'. In other words the theory holds that some progeny will always be better 'adapted' to their environment than others, and will therefore survive, multiply and perpetuate their characteristics. This is the 'preservation of favoured races in the struggle for life' of the full title of the *Origin*, an idea expressed more succinctly in the phrase 'the survival of the fittest'.

Darwin was not only aware that individual progeny vary in different degrees from their parents and that brothers and sisters may be more alike or more dissimilar, but he knew also that there were occasionally produced individuals so unlike their parents or siblings that they could scarcely be described in the same terms. With regard to these 'sports', later to be more widely called 'mutations', he found himself in a difficulty because such marked and sudden appearances of something new could easily be interpreted as at least akin to 'special creation'. Whether, as seems likely, it was for this reason or not, he concentrated his attention on small, or 'infinitesimal', variations, paying comparatively little attention to mutations, and the Theory of Natural Selection lays particular stress on the former. Curiously enough neither Darwin, nor those opposed to his views, seem to have realised that small variations could just as easily be attributed to supernatural agencies as could large ones, but this apparent inconsistency may seem more obvious now in the light of modern thought than it did to the intellectual world of mid-Victorian times.

Another matter that can be argued is whether Darwin, and more particularly Wallace, whose experience of equatorial life was greater, gained the idea of natural selection from what they had observed in feral nature or

47

whether they based it on the known facts of selective breeding, but certainly the *Origin* contains a great deal about the latter. This was only to be expected because its author realized that if the views he was propounding, so opposed as they were to the prevailing doctrine of special creation, were to win acceptance, he would have to express them in phrases and analogies familiar to the reading public. This he did by focusing attention on variation and illustrating his remarks by references to domestic plants and animals, and it is this recognition of variation as the key to that change with time which is an essential element in biological evolution that may rightly be regarded as his greatest contribution to biological thought. It should also be remembered that the *Origin* appeared at what is commonly, but not altogether accurately, called the 'psychological moment', when a gradual and growing shift of opinion away from the idea of special creation made the public mind particularly receptive to the views expressed in it. Even so the book had a mixed reception for some years but before the death of its author it had become the corner-stone of the edifice of evolutionary literature.

So it was that, although it took some time, the Theory of Natural Selection eventually became accepted by the scientific world and has ever since been the best-known theory about organic evolution. There were, and still are, critics of the theory, and from its earliest days some of the arguments against it have never been adequately answered or rebutted, but they have become submerged beneath the weight of the approbation that the Theory has, sometimes rather uncritically, received, and it is therefore very desirable to enquire into it more closely here.

The idea of 'natural selection' rests on two propositions: first that the populations of assemblages (that is to say the numbers of individuals in each) are, under feral conditions, more or less stable, fluctuating but little with the short-range passage of time; and second, that in nearly all kinds of plants, and in the great majority of animals, potential progeny, or 'propagules' as it is convenient to call them, are produced in numbers vastly in excess of those needed to maintain the populations at about the same level from generation to generation.

Of these two propositions the first calls for the more careful consideration because there are circumstances which seem, at first sight to be inconsistent with it. It is, for instance, a commonplace that there are 'good' and 'bad' seasons for many plants and animals, the two adjectives being used to express more or fewer extant individuals, and it is often all too obvious that the individuals of many assemblages occur, on occasion, in such large numbers that they become a nuisance or danger to man and so acquire the epithet of 'pest', and it is easy to regard examples of this as evidence against the proposition. However, closer examination of such cases reveals that in most of them, and certainly in the most conspicuous, the opportunity for inordinate multiplication is presented by the direct or indirect influences of man exerted, at least in those parts of the world which he has himself most densely populated, in climatic conditions which vary considerably from year to year. Man's primary concern is to feed himself and to do this he has developed a complex system of agriculture of which the most noteworthy feature is the production of large and locally concentrated stores of food – his crops, and it is the existence of these which encourages the undue multiplication and spread of 'pests'.

More often than not climatic and other environmental conditions prevent these abnormal increases from becoming a menace, but when great resources and unusually favourable climatic conditions coincide the result may be what is commonly called a 'population explosion'. The resultant 'plagues', as they are generally termed, and the evidence they seem to afford, provides one of the chief arguments against the proposition of population stability, but this is misleading.

When these plagues occur man naturally does what he can to control them, and advances in agricultural

49

practice largely take the form of improved methods of doing this, so that though they may be serious their effects are comparatively short-lived and the numbers subside again fairly quickly. Even if man took no remedial steps this would happen before long because the augmented populations would soon exhaust the supplies available to them. This in itself is sufficient to dispose of any contention that explosions of this sort are evidence against the proposition of population stability but it is circumstantial evidence and would be strengthened by more direct testimony derived from feral nature. Unfortunately this is difficult to come by because the influences of man are now so widespread and far-reaching that it is hard to find any natural society of plants or animals which is entirely free from its effects. Moreover, if such could be found it would be impossible to study it for any useful length of time without importing the very human influences which would vitiate the results of any observations made. Hence a belief in what is here described as population stability is based chiefly on the observation that although there may be wide fluctuations in individual numbers over such short periods as runs of a few years these are but temporary variations which tend to cancel themselves out in due course and which cannot therefore be regarded as evidence against the opinion that populations are, in general terms, stable, and liable to significant change only slowly and over comparatively long periods.

The evolutionary aspects of the problems presented by pests and plagues are of great interest. Man's ever-increasing capabilities have enabled him to populate the world in many ways that would otherwise have been impossible, and in the process he has taken with him, not only the comparatively few kinds of plants and animals of special value to himself, but many others that have accompanied him without his direct knowledge. He has thus disrupted the patterns of natural distribution, not only of the plants and animals that he needs but also those of many others, and he has gone far towards upsetting the true balance of nature in nearly all those parts of the world that he has occupied. He has done this principally by his cultivation of crops and breeding of stock because, as has been mentioned, the effect of this is to provide concentrated local supplies of certain food-stuffs which would otherwise not be obtainable in necessary quantity, and these offer great attractions and opportunity for many other plants and animals. This distorts the whole ecological balance because the restrictions of food limitation are often to a large degree removed; increased multiplication is facilitated; and, above all, the destruction of excess propagules is reduced. Familiar examples of this range from the introduction of prickly-pear cacti into the more arid regions of the Old World, and of the rabbit into various temperate countries, to the more intricate matter of epidemic disease, but all go to show the fine balance of feral nature; how easily man can disturb it; and how misleading, therefore, it is to apply human standards of struggle and destruction to virgin nature.

With regard to the significance of man's crops in the spread and multiplication of pests and plagues, and in relation to the general argument concerning population stability, it should not be forgotten that, in terms of cosmic time, it is but yesterday that these became an important element in the general ecological picture, and that therefore their value as evidence of any state of affairs that may have existed during the long ages before man made his effective appearance must certainly not be exaggerated.

Finally with regard to the evidence relating to this first proposition there may be added the strong indication given by the overall impression, which a wide practical knowledge of biology certainly conveys, that the enormous number of different assemblages which make up the plant and animal kingdoms could not exist, and continue to exist, if there were not some general environmental control of populations. This is the most important indication of all and that it is difficult to demonstrate directly should not be allowed to diminish its importance.

The second proposition, that nearly all kinds of plants and many kinds of animals produce propagules in numbers greatly in excess of those necessary merely to maintain the population of their various assemblages, is easier to deal with because it is based on clearly observable and calculable facts about which there can be little dispute.

In order that the numbers of individuals in assemblages shall be maintained from generation to generation it is only required that each and every parental pair shall, before death overtakes them, produce two surviving and fertile progeny, but common observation has made it abundantly clear that propagules are generally produced in numbers far in excess of this requirement.

In the plant world all the major types show this over-production and, indeed, some of the most extreme examples are found among highly specialized flowering-plants, notably the orchids, but in the animal world the most conspicuous instances are seen in some aquatic groups, especially the fishes, in some of which millions of eggs are produced by each female every year. In land animals generally things are rather different. In some of the insects, especially those that fly, geat numbers of eggs are commonly laid, but in many other groups production is far less. The culmination of this is to be seen in the mammals and particularly in those that have long gestation periods and in which multiple births are unusual. This is certainly associated with the evolutionary development of vivipary and is yet another aspect of the exploitation of the subaerial environment, a subject which is the main theme of the next part of this book. Yet despite this great over-production of propagules population numbers do not fluctuate in any corresponding degree and hence it must be accepted that nearly all the progeny produced in such large numbers fail to survive into reproductive maturity and consequently that there is something describable as a mass destruction of them. The agencies of this mass destruction are innumerable but the fundamental cause of it lies in one of the most important, but often inadequately understood, principles of biology.

This is that feral living nature wholly unaffected by human influences is in a state of equilibrium which changes only very slowly. To this condition of balance every different kind of plant and animal subscribes in such a way and in such measure that each finds, and keeps, its appropriate place within the framework of supply and demand and does not draw inordinately, or at the expense of others, on the total resources of the environment, which resources themselves impose strict controls on the numbers of individuals that can be supported. This subscription mainly takes the form of the destruction of excess propagules just described because this overproduction provides not only an immense supply of food but also a reserve from which short-term fluctuations in numbers can be met.

The continued existence of a huge number of different assemblages, of all degrees of potentiality to predators, is sufficient evidence of this remarkable balance and every side of living nature provides examples of it, one of the simplest being that of predatory animals such as birds of prey and carnivores, where predators and prey keep a balance, as indeed they must if both are to survive. It is the failure to understand this basic commensalism and communion in feral nature that has led in the past to the currency of the unfortunate phrase 'nature red in tooth and claw' which has been made the justification or excuse for many kinds of anti-social and attitudes.

To stress the general point even more cogently it may be repeated that if an assemblage is to maintain its population of individuals at the same level from generation to generation no more is required than that each parental pair shall, during their lives, produce two offspring which shall themselves do likewise in due course, and so on. This means that in plants and animals producing thousands and even millions of offspring during the reproductive activity of parents, only about two of these survive, and to suggest that these two are, in any significant way, superior in evolutionary potential to the vast numbers that succumb, and survive for this reason, is manifestly unrealistic. This is so patent as scarcely to need further comment, but it is worth while to point out also that the survivors in this kind of situation can be regarded as 'fittest' only on the circular argument that they have 'weathered the storm': they are said to have survived because they are the fittest and they are regarded as the fittest because they have survived. It is unfortunate that this imperfect concept of the 'survival of the fittest' should have come to occupy so prominent a place in Darwinian theory, because it has led to repercussions far beyond the bounds of scientific biology.

It is not easy to find a simple analogy which gives a true picture of the circumstances just outlined because human valuations intrude from all directions, but the sowing of seed in agriculture, which is not so very different from the dissemination that takes place in nature, is one of the most useful. Here the chief distinction between artifice and nature is that in the former the terrain over which the seed is scattered has been brought, as far as feasible, into a uniform condition calculated to be particularly suitable for the kinds of plants whose seeds are being sown; while in the latter the terrain is not deliberately prepared by human hands or machines and is consequently, in the great majority of cases, and above all in virgin nature, far more diverse in character. Yet even in the former case allowance has to be made for a wastage of seed during and after sowing, and it certainly cannot be argued validly that the seeds which fail to grow do so because they are *in themselves* the least fitted to do so, nor, to reverse the argument, can it be held that those which grow successfully do so because they possess certain innate characteristics of special value to them in the circumstances; rather they owe most to the chances of fortune. How much less can the same argument be applied to situations which are uninfluenced by man, and where, in general, conditions are quite different?

It is useful at this point to draw a distinction between propagules (usually spores, eggs or seeds) and the sporelings, larvae or seedlings to which they give rise. This distinction is least evident among marine forms of life, notably the seaweeds and many invertebrate animals, where both stages are equally vulnerable and are passed in the same environment, and it is much clearer in many land organisms in which eggs and hatched young are often very different in terms of vulnerability.It is especially plain in seed-plants where seeds and seedlings differ widely in their relationship to the many hazards of their surroundings. This particular aspect of life on land reaches its climax with the development, in sexual reproduction, of vivipary: the condition in which the female gametes do not leave the parent body and the embryos do not emerge until they have passed through the initial stages of their lives. All these examples show that the environment may present different challenges to successive life-stages, and it by no means follows that because the qualities of an individual at one stage in its life are advantageous to it those of a later stage will be the same.

Probably few will deny the value of the *idea* of 'natural selection' or doubt that among any collection of individuals some will, because of the combinations of their characteristics, be more likely, in the circumstance prevailing at the time, to survive and propagate than others, but it is highly debatable whether these are in fact the individuals which *do* reach reproductive maturity; and, even if they do, whether the characteristics that have contributed to that survival do, in the course of successive generations, become cumulative and so eventually produce descendents which can be described, in current phrase, as constituting a 'new species'.

This last point is the crux of the whole matter. Every generation a new population of individuals is produced which is exposed to the contemporary conditions of its environment, but this environment itself changes with time and therefore the individuals of later generations will not be exposed to exactly the same conditions as were their forerunners. In these circumstances there is no good reason for supposing that any characteristics which may have helped individuals of any particular generation to survive and multiply will be equally useful to the individuals of subsequent generations, and without this *continuity of effectiveness* what is generally called 'natural selection' will not lead to that sustained directional change which is the essence of organic evolution.

Such considerations lead to the conclusion that the process that has been given the somewhat prepossessive name of 'natural selection' is more truly to be regarded as a more or less random sifting in which selection, in the true meaning of the word, has no place. If this is so then the problem for the biologist is not whether 'natural selection' occurs but whether the sifting which is its more proper description is, and has been, a major factor in evolution.

Here much depends on what is meant by a 'factor' in evolution. If it means an influence that has determined the course of evolution then the process, which may be called 'selection' or 'sifting', is clearly such because it determines that not *all* but only *some* individuals survive and that the qualities possessed by those which do not so survive remain unexpressed; although, bearing in mind what a very small proportion of offspring reach maturity, this must be looked upon rather as a limiting factor. If, however, a factor is taken to mean a major *cause* of evolution then neither sifting nor selection can be rated as more than a subsidiary effect. The real factor is that, since no two propagules or individuals are alike, there must inevitably be variation between generations, and it is this which is the real basis of organic evolution. That evolution has followed the course it has must primarily be due to this, but why it has expressed itself more particularly as it has is certainly due to the operation of a sifting process, because this *precludes the expression of the qualities of all those offspring which do not happen to reach reproductive maturity*, and these lost qualities may well include some of great potentiality.

Another point which leads one to doubt the appropriateness, in the study of feral nature, of any process worthy of the name of 'natural selection' is that the phrase itself reflects the common fallacy involved in the application of the ideas of 'better' and 'worse'. This matter has already been discussed with particular reference to classification but it has a much wider general significance, and it may well be repeated here that it is impossible to divorce from the word 'selection' the notion of choice by preference; that preference in turn can have real meaning only in terms of relative merit; and that merit can only be estimated in terms of the attainment of some end foreseen by the human mind.

Useful as the analogy described on p. 54 may be in commenting on the Theory of Natural Selection it is of even greater value in the way it leads directly to another aspect of the matter which has been consistently underrated. A little later on stress will, quite appropriately, be laid on the essential similarities between the plant and animal kingdoms but from the point of view of 'natural selection' there is one quite fundamental distinction between them. Plants, except for some of the simplest of all, are passive and immobile, but animals, again with a few exceptions, are active and mobile. This means that the former are 'rooted to the spot' in a very real sense and are dependent on their immediate surroundings for their supplies, but that the latter can, either entirely by instinct or by the exercise of some degree of volition, change their locations. They are, in short, capable of taking evasive action against some at least of the dangers that threaten them. To give but one simple example among countless others, animals are able to stave off partial or total starvation by moving to 'greener pastures' or, if they are predators, by changing their prey or their hunting ranges. They are able to vary their environment by movement. Plants cannot do this because they are 'bound to the soil' in a particularly intimate way. Their environment is that in which they find themselves as a result of passive distribution containing a large element of chance, and even if their surroundings alter during the course of their lives they are incapable of accommodating themselves to the change.

The impact of this distinction on the Theory of Natural Selection is profound, and offers an enticing field of further enquiry, but whatever the outcome of this might be it is clear that the idea of 'natural selection' as usually described and illustrated cannot apply equally to both the plant and animal kingdoms and that the features of it that make it an attractive proposition to some zoologists are just those which make it unacceptable to botanists. No doubt it can be argued that the idea of natural selection emanated from the minds of zoologists and that it is not meant necessarily to apply to plants, but the inescapable corollary to this view is that the evolution of plants is, and has been, different in kind from the evolution of animals, and this all the available evidence refutes.

But perhaps the most cogent comment on the Theory of Natural Selection as understood in Darwinian terms is that it puts the cart before the horse. For the moment this point may be put in this way. Given variation, and especially that between parents and offspring, in a gradually and continuously changing environment such as the world affords, the kind of change that is called evolution is inevitable, and within this something akin to what is mistakenly called 'natural selection' is equally inevitable, because as a result of this variation some individuals will fit the conditions of their environment more appropriately than will others and will, if there is powerful competition, be *more likely* to survive, but this is an *effect* and not a *cause* of evolution. The more real problem is why evolution has been the orderly progression that all evidence indicates, and here again the idea of 'natural selection', though it may be an observed consequence, cannot be accepted as a *vera causa*.

These comments are not to be regarded as a criticism of the Theory of Natural Selection, but rather as a corrective. Without a comprehensive theory as under-standable as this to the lay mind the great progress that the study of evolution has made in the last century or so would scarcely have been possible, and the theory closely fitted the requirements, and also the rather chauvinistic mood, of the time. Since then experience, often bitter, has shown that things are not quite so simple as the theory might suggest and some rationalisation of it has long been due. On one hand the theory has had an unfortunate influence on the course of certain political trends: on another it is too 'comfortable' a theory, and at first sounds so satisfying that the urge to find something even better tends to be quelled. Last, but by no means least, it has become unduly identified with the story and personality of the better known of its distinguished originators and on this account has attracted so much respect that it calls for some determination to point out its limitations.

None of these comments however afford a sufficient explanation of why the Darwinian idea of natural selection has led to so much confused thought, but some of the difficulty can best be understood by reconsidering the situation with regard to evolution as it was in the years immediately preceding the publication of the *Origin*. At this time, in the 1840s and 1850s, the idea that the different kinds of plants and animals had been 'specially' created and were immutable was still very widely held, but the opposite opinion that they were *not* immutable but were the products of change over long periods in the past was held by many, chiefly scientists, and this opinion was spreading, at least below the surface. This was principally due to the growing evidence afforded on the one hand by the widening application of methods of selective breeding and on the other by the rapidly increasing knowledge of what is now called the fossil record, which showed beyond doubt that the world had, in times gone by, been inhabited by a succession of different, and now unrepresented, kinds of flora and fauna. These pointed clearly to the conclusion that the living world was changing, could be changed, and had been changing for a vast period of time, though it may be doubted whether the word evolution in anything like its present meaning was at all current. At all events Darwin's primary object in writing the *Origin* was to demonstrate the mutability of plants and animals and for this purpose he made great use of the facts of selective breeding; but, more importantly, he tried to propound a reasonable and easily understood explanation of how this mutability comes about in nature, and of its consequences. It is particularly to be remembered that there was at that time no vocabulary of the nascent subject of evolution and that in order to make his meaning clear to the scientifically uninformed he was obliged to write in terms of everyday life using familiar words and phrases to describe what were, in fact, unfamiliar theses. It is this inapplicability of certain words to certain ideas which eventually led, when the concept of evolution as a complicated scientific process had become widely accepted, to misunderstandings which still persist.

These misunderstandings derive particularly from the employment of several words which it was reasonable enough to use at the time in what was, after all, an illustrative fashion, but which it has since become common practice to use more precisely. Moreover, these words were, unavoidably in the existing circumstances, words which either described human actions or reactions or which were reflections of human sensory perception, and this, as is repeatedly demonstrated in these pages, imported into the study of evolution a sense of human valuation or, to use a clumsy but apparently irreplaceable word, an anthropomorphic, background which, mainly through its later exploitation, gives a falsely humanised picture of many aspects of the non-human living world from which it has subsequently proved most difficult to escape.

The most prominent example of this is the use of the word 'selection' in the title of Darwin's theory. Selection means, by definition, deliberate choice by preference and this can be exercised only by a discriminatory intelligence for the existence of which there is, in the non-human world, no evidence. The use of the word is thus prepossessive, begging the question to which it is applied. Another conspicuous instance of verbal misapplication is seen in the use of the word 'mimicry', and this particular case is discussed on p. 71. More troublesome and confusing than either of these is the case of 'adaptation', a word prevalent in many evolutionary contexts. In ordinary language adaptation means the purposive modification of one thing or subject so as to make it more closely resemble another thing or subject in form or function or in both. The use of the word in a biological connection therefore inevitably suggests that such a process occurs in nature, for otherwise there would be no point in using the word at all, but how far this is true has still to be demonstrated. The first use of the word in a biological sense is usually attributed to the *Origin*, where 'adaptations' and 'co-adaptations' are referred to, and it is from this that the now familiar phrase 'adaptation to the environment' comes. The difficulty here is that the definition of the word adaptation just given presupposes that one of the items in such a process is undergoing change while the other is relatively static. If it were not so, and if both items were simultaneously changing at the same rate it is hard to see how there could be anything describable as adaptation. Yet this is exactly the situation that pertains in nature, because the environment, since it is part and parcel of a ceaselessly changing planet, is, like every thing else itself subject to change. It is also hard to understand precisely what is meant by 'the environment'. In common phrase the word describes the *milieu* in which each and every organism finds itself, but this *milieu* varies greatly in kind. As far as plants are concerned it may be almost totally mineral in constitution but in animals it must consist in some degree, and often wholly, of members of the plant world.

Nor are the many attempts made by biologists to define the concept of adaptation more exactly of much help, and many commentators on the subject seem to be reduced to regarding 'adaptation' as no more than one way of expressing the organism's ability to live in the surroundings in which, through a host of circumstances, it finds itself. Here again there is the kind of circular argument that was noticed in connection with 'natural selection'. An organism continues to live because it is adapted to its environment, and it is regarded as adapted to its environment because it continues to live. Still another point is that the word adaptation, in its ordinary application implies an element of intent directed towards an identifiable goal, and this, as is discussed elsewhere in these pages is an unsuitable basis for the study of evolution. It is this inherent implication of design that makes 'Darwinian' adaptation so hard to grasp. Nor is there any real need for such a concept. All available evidence goes to show that the mineral world, the plant world, and the animal world are all in a state of constant change. They are all evolving according to their respective natures and one of the principal results of this evolution is that living organisms arrange themselves according to their mutual ecological appropriatenesses, and failure to find a place in this equilibrium may be a lethal factor.

There is another matter closely associated with the Theory of Natural Selection which calls for brief mention at this point. Darwin, in his chapter on 'natural selection' in the *Origin*, incorporated shortly and almost parenthetically, a second principle or hypothesis which he called Sexual Selection and which is intended to explain, within the context of 'natural selection', the occurrence in various animal species of secondary characters distinguishing the two sexes. Especially familiar among these are the aggressive features (antlers, horns, tusks, spurs and so on) in some of the vertebrates, especially mammals and birds, with which the males fight between themselves for the possession or protection of the females: the brilliant colours of many males compared with the females, most notable in birds: and the various kinds of sexual display, including bird song, by which the males appear to court the females.

As regards the first of these it is not difficult to maintain that the weaker males will be discouraged or eliminated in the competition and that the race will therefore be selectively improved by the repeated infusion of blood from the stronger, but even here there are many possible factors which would make a long-term result of this sort uncertain even if it were biologically beneficial, and these misgivings are much stronger in the other two examples. To put all the pros and cons of 'sexual selection' into words would need more space than is available here, but two general comments may enable the reader to make some judgment of the probabilities for himself.

Differences of appearance between males and females of the same assemblages ('sexual dimorphism' as it is called) is in fact widespread in the biological world not only in the instances just given but also in many invertebrates, particularly the arthropods, and in many plants of almost all groups, so that any general explanation of it must be applicable much more widely than merely to those vertebrates most often quoted.

The other, and more profound, difficulty in the theory is that it involves the emotions of animals, and these can, in fact, be judged only from our own human experience. This subject, which presumably must take into account some degree of aesthetic preference and appreciation, provides very thin ice indeed on to which only the more rash would venture, and this book is certainly not the place for such an undertaking. With this comment it is best here to let Darwin's Theory of Sexual Selection rest though it should not be wholly neglected or overlooked because further and more penetrating investigation may well lead in valuable, and perhaps unexpected, directions, but it is a just assessment to say that the theory, which finds mention in many books on evolution, does not now receive the regard which its author considered it to merit.

Both before and since Darwin's time many other ideas have been propounded to account for evolution, and two of these link so closely with the notion of 'natural selection' that they must be referred to briefly here. These are 'use and disuse' and 'the inheritance of acquired characters'; ideas which no doubt originated in man's mind long ago, but which are generally regarded as having been first given precise expression in the writings of the French biologist Jean Baptiste Lamarck who lived from 1744 to 1829, and they make up the essential parts of what is usually called Lamarckism.

The concept of 'use and disuse' embodies the idea that the more an organ of the body is used, or its functions exercised, the more these become accentuated and developed, so that the features concerned will gradually become altered, producing one kind at least of the 'change with time' which is the chief expression of biological evolution. Contrariwise the less organs and functions are exercised the more they will tend to diminish or atrophy. Clearly a cause for this must be postulated and the Theory of Use and Disuse supposes that it is activated by changes in the environment and mode of life of the organisms concerned, to which it is the response.

A little thought will make it plain that the value of this idea from the point of view of evolution depends on whether or not the changes of character which may come about during the life-time of the individual under the influence of an altering environment are, in fact, transmitted to their offspring, and difference of opinion on this point has led to intense controversy on more than one occasion. This problem of the 'inheritance of acquired characters' has repeatedly attracted attention, partly because it seems to offer a possible alternative to the Theory of Natural Selection and partly because it seems to afford an opportunity for studying the reactions of animals (like 'use and disuse' it is essentially a zoological concept) to their environment by actual experiment. Many such enquiries have indeed been carried out but it has never been possible to regard the result of any of them as more than inconclusive. This is one reason why the subject of the 'inheritance of acquired characters' is one of the most contentious branches of the study of evolution, though it is not at all clear why this should be.

All new characters, whatever they may be, or however they may arise, must necessarily make their first appearance during the life-span of one or more individuals. The phrase 'inheritance of acquired characters' is based on the supposition that while some new characters are the expressions of genetic changes in the parents of the individuals showing them for the first time, others are the direct expression of changes brought about by the influence of the environment *during* the life of the individual concerned. It is to the latter that the phrase 'inheritance of acquired characters' is applied and the problem is to determine whether these 'acquired' characters are inherited; i.e. passed on from generation to generation by parents. If they are then it can only be through some modification in the genetic apparatus of these parents: if they are not so inherited then they can have but little, or no, influence on evolution. The question of whether 'acquired' characters are inherited or not is therefore an important one, at least in theory, and the failure to reach conclusion on the matter suggests that there may be some misappreciation of the situation, though just what this may be is difficult to discern. Every emergent individual contains the potentialities conferred upon it by the genetic system of its parents, and the question is therefore whether this system can be influenced by the environment during the lifetime of the individual in such a way that the change resulting from that influence can express itself genetically in a parent before a new generation is conceived. If they can be so expressed then what are called 'acquired' characters are, or may be, inherited, and the process must be reckoned as an important factor in evolution.

It is from the Theory of Natural Selection that there also stem the highly controversial subjects that are commonly called 'protective resemblance'; 'camouflage'; 'mimicry'; and 'warning colouration'. Examples of these have been known for a long time, and in early days were accepted as wonders of nature without much speculation about their significance, but the idea of 'natural selection' much quickened interest in them because this theory seemed to provide an opportunity for the kind of teleological explanation of them which once seemed so appropriate and satisfying, and since that time they have been the subjects of much rather confused and confusing argument. Many attempts have been made to substantiate these teleological opinions by experiment, but the deeper investigation of individual instances has nearly always encountered contrary evidences that throw doubt on the validity of the assumptions which these experiments seek to confirm, and it is true to say that, despite the number, ingenuity and perseverance of these studies there has never emerged from them any easily comprehensible hypothesis that will explain all the facts. This suggests that the premises on which these enquiries have been conducted are more or less false, and it is not difficult to see where some of this falsity lies.

Unfortunately few aspects of evolution are expressed by such an array of isolated and detailed facts with so little obvious relationship to one another as are these subjects, and because of this few other aspects have become so distorted by argument from the particular to the general. Moreover the facts themselves are so miscellaneous and of such unequal significance that it is difficult to knit them together into any sort of coherent account that is more than a mere recital of them.

Nearly all the cases to which the four terms just referred to apply are more properly described as examples of *superficial resemblance*. These two words are not entirely appropriate but they are commonly and usefully accepted to mean resemblances between individuals greater than might be expected from the degree of relationship between them, that is to say from their genealogical affinities. Similarity of this kind is far more widespread in the living world than is generally realised, although the comparatively few most striking examples of it in animals are familiar enough. Among plants superficial resemblances is so usual that it tends to escape notice especially where it concerns the vegetative parts but it is nevertheless very real, as two obvious examples taken from the Angiosperms testify. Growth-form in these plants ranges from ephemeral annual herbs to giant forest trees but the number of different forms is strictly limited and it is a commonplace that some of them occur over and over again in families which, at least in terms of formal classification, have no close connection. By contrast some growth-forms, e.g. the 'ericoid' habit, are comparatively uncommon and therefore attract more attention. Superficial resemblance is even more remarkable among the foliage organs. These consist of leaves, which in Angiosperms are organs *sui generis* without exact parallel elsewhere in the plant kingdom, and here, just as with growth-form, the number of different designs is strictly limited and almost as strictly unsegregated. A comparatively small number of basic leaf-designs are repeated throughout the group as a whole and the character of leaf-form is therefore not made a criterion of genealogical relationship. Only very rarely do unique leaf-shapes occur, a point cogently stressed by the familiar difficulty of identifying fossil Angiosperms from their detached leaves.

To a lesser degree other great groups of the plant kingdom show the same kind of superficial resemblance within themselves, as is particularly noteworthy in the marine algae, where plant form in its simplified expression of the thallus is similarly repetitive. Less obvious but of even greater interest are the many superficial resemblances between individual flowers or whole inflorescences of Angiosperms and these are also the result of the frequent repetition of certain basic plans and patterns in the different groups of these plants.

It thus appears that superficial resemblance is by no means confined to animals and is certainly more generalised in plants, and easier to understand there, because it is not complicated by the volition and great variation in life-pattern shown by animals, but for this very reason these plant resemblances have never been credited with such values as are inherent in the phrases 'protective resemblance', 'camouflage', 'mimicry' and 'warning colouration'.

In view of this it is relevant to stress that the instances of superficial resemblance in animals to which attention has particularly and sometimes inordinately, been directed are, in fact and comparison, very few and found mainly in certain groups. This may be demonstrated by arranging the various examples in a way that conveys the relative frequency of each. Botanical, zoological and mineral components all contribute to superficial resemblance of one kind or another, and it is revealing to note how widely or otherwise various combinations of the three occur. The most widespread kind of resemblance is that between different plants, and corresponding instances among animals are not only fewer but are also strongly segregated taxonomically, the most conspicuous being scanty in number and especially to be seen in the great group of the insects. Resemblances between animals and their immediate surroundings, which latter usually comprehend both plant and mineral components, are more frequent though often less definable, and it is with these that the idea of protection and camouflage is particularly associated. Moreover these are found in many animal groups, including all the sections of the Vertebrata. Much less frequent are three other combinations, namely resemblance between plants and their habitats, which in its most striking expression is largely confined to some of the desert plants of southern Africa, and to some of the lichens and algae; between plants and animals, such as is almost invariably instanced by reference to the orchid genus *Ophrys*; and between animals and plants. These last particularly seem to express the idea of camouflage and are well illustrated by various leaf- and stick-insects and by certain Lepidoptera. A survey of these differing examples clearly suggests that superficial resemblance is likely to result from the repetition of certain particular combinations of form and colour and need not necessarily be purposive.

In many cases of 'protective resemblance' the whole animal, immature or adult, blends into its background because it has, in man's appreciation, a colour pattern that appears calculated to reduce the likelihood that the animal will be detected by a predator and it is understandable that it is in some of these cases in particular that there is the clearest suggestion of 'natural selection', on the grounds that the more perfect the blending with the background the more likely that the animal kind concerned will survive, thus putting a premium on increasing degrees of resemblance. There are, however, one or two profound objections to this view. Why, if it is true that these resemblances are advantageous, are the assemblages exhibiting it not more common and with larger populations, and why, on a wider scale, is this sort of camouflage not more widely seen in the animals kingdom? More particularly, how far can the perceptual appreciations of man be applied to other animals? In many of the cases under discussion the camouflage does not result solely from a similarity of pattern and colour but depends much on the deeply implanted instinct of animals to remain motionless in danger, any movement on their part being likely to destroy at once this camouflage and to attract attention.

This last point leads directly to what is certainly the most important aspect of the whole problem of superficial resemblance in animals, namely that it depends so much on perception and, in particular, on the one perceptual sense, sight, which is generally held to be more highly developed in man than in any other animals. This pinpoints in a peculiarly inescapable way one of the chief general obstacles to a better understanding of evolution; namely the constant tendency to interpret the reactions and behaviour of other animals in terms of those of the human kind. That the weakness of this anthropomorphic approach is not more readily apparent is surprising because the similarities between the plant and animals kingdoms greatly outweigh the differences between them and much that goes for the one also goes for the other, yet the suggestion that such explanations of superficial resemblance as go by the names of 'protective resemblance' and 'mimicry' are applicable, in teleological terms, to plants can hardly be sustained.

But to return to the subject of sight, about which there is much more to be said*. Much of its value in explanation of many of the similarities in the animals kingdom rests on the assumption that the faculty of vision in other animals is at least closely similar to that of man, but to this view much evidence, both ophthalmological and behavioural, is opposed. Visual power in animals, and especially that of distinguishing colour, is extremely varied in both scope and value. Often some members of a particular group appear to have colour vision, while other members which *a priori* might be thought equally likely to possess it, are colour blind. Nor do all animals rely equally on the visual sense; many are nocturnal in their habits or live permanently where no light penetrates. Many other instances could be quoted but the whole issue is summed up in the opinion, widely held by those most conversant with the facts and problems of vision, that the only animals in which powers of colour vision compare with those of man are certain lizards; some of the diurnal carnivorous mammals; some ungulates; the Primates; and, above all, the birds.

The case of the birds is especially noteworthy because, apart from anything else, they are the wild animals in which everyday observation of behaviour is easiest, a fact that accounts for much of the current interest in bird-watching. Not unnaturally in the circumstances it is here that anthropomorphic interpretations seem most apt but it is also true to say that attempts to establish them on a firmer basis by experiment nearly always eventually meet

---

* Much of the information in this and the next paragraph is derived from Duke-Elder, Sir S.: *System of Ophthalmology*, vol. I, London, 1958.

with frustration. This can be due to many causes but fundamentally it is because birds rely far less on the sense of sight than man does, and only in some of them is that reliance so great as to make ideas like 'protective resemblance' and the rest relevant. For instance it is patent that in many birds of prey the visual reactor mechanism owes much more to the perception of luminosity and movement than to actual acuity of colour vision.

Another aspect of 'protective resemblance' and camouflage that must be referred to briefly is that of melanism. This is defined as a more than normal development of 'dark colouring matter', and gives name to the adjective 'melanic', meaning black or dark in pigmentation. As happens so often when superficial resemblance is under discussion, the facts of melanism have become confused by an excessive concentration of interest on one or two particular examples of it, notable among them being the case of certain moths living within industrial areas, which are reputed to have become melanic in response to the partly artificial conditions of their environment, a response which, in local conditions, gives them a desired degree of protection from their predators which they would not otherwise possess. The weakness of this somewhat facile explanation is that it fails to take account of at least four circumstances. Melanism, in the form of the occasional appearance of dark or black individuals in assemblages normally otherwise coloured, has long been familiar and is common in both plants and animals, a point sufficiently illustrated by the old phrase 'the black sheep of the family'. Again, much, and possibly all, melanism is known to be controlled by genetic factors, and that these are directly controlled by prevailing industrial conditions has not been satisfactorily established. Another point is that melanism more often than not occurs in animals, chiefly Lepidopterous insects, whose natural distributions have little or no direct association with areas of high industrialism. Lastly it may be borne in mind that melanism in industrial areas may be quite rationally

71

accounted for by supposing that larvae and other immature stages acquire their melanism by feeding on in-
dustrially polluted food.

Finally with regard to 'protective resemblance' by far
the most significant objection to the usual teleological
interpretation of the facts is that, by implication, it
attributes to other animals the notoriously imperfect
perceptual powers of man. Can it really be believed that
non-human animals have no better sense-values in hearing, smelling, feeling and tasting than man has, and can it
really be supposed that predators whose lives largely
depend on their ability to detect so called camouflage
cannot penetrate it better than man, who has but curi-
ousity in it? All human experience with many other
animals indicates that their sensory perceptions are, in
many respects, greatly superior, and to suppose otherwise
is a poor compliment indeed to the obvious abilities of these
animals to fend for and protect themselves in ways which
men cannot match. All this, moreover, beside the wide
array of evidence for the view that many animals, notably
mammals, pelagic fish and insects, are clearly able to
employ additional senses of which the human mind knows
little or nothing.

The inadequacy of the anthropomorphic approach is
even more apparent in the phenomenon known as 'mimicry' because this description prostitutes the very word
itself by subtly inferring a change of meaning. In its
original contexts the word mimicry means 'imitation with
intent' and this meaning is clearly inapplicable to superficial resemblance in plants and in non-human animals.
This difficulty is sometimes countered by maintaining that
the word mimicry is used only in a neutral sense of
resemblance but it does not seem to be realised that this
very argument, if it deserves to be so called, makes pointless
the use of the word in this connection. Deliberate imitation, in the proper sense of the phrase, is a product of
human intelligence, and to suggest, as the use of the word

mimicry does, that animals other than man have this ability is unwarranted.

In more practical terms the most important fact about 'mimicry' is that although intense interest has been aroused by some of the most conspicuous examples of it, it is very uncommon and in its most complete expression is almost confined to a small number of insects. This point is important because it is the basis of one of the most cogent criticisms of many ideas associated with the more conspicuous and less common kinds of superficial resemblance which takes the form of the question why, if these similarities have the selective and survival values so often attributed to them, are they so singularly infrequent, and why are the assemblages showing them are not more 'successful' and more prominent among their fellows? The proportionate occurrence of these resemblances certainly suggests that far from being of genealogical value they are in fact examples of 'blind alley' evolution and may not unreasonably be described as aberrations.

Doubts about the validity of the idea of 'mimicry' are reflected in one of the best known and often quoted expressions of it, namely the recognition of two kinds of 'mimicry' called, respectively, Batesian and Muellerian. Batesian mimicry is the name given to those cases in which, to quote a common wording, 'a defenceless organism bears a close resemblance to a noxious and conspicuous one'. Muellerian mimicry is described as the 'tendency for noxious species to resemble one another'. Both of the definitions rest on what is purely an assumption that some kinds of animals are noxious while others are not, though it is never satisfactorily explained how it is known that they are noxious nor in what way and to what other animals they are noxious. It is not even clear whether this attribution is supported by the sensory reactions of man himself and, as has already been pointed out, this is a subject that impinges dangerously on the sphere of human aesthetic appreciation. With this comment, which many will agree is sufficient to discredit, or at least raise doubt about much of what has been written about 'mimicry', the subject may be left.

What commonly goes by the name of 'warning colouration' is, in many respects, the opposite of 'protective resemblance' but the facts about it can be stated just as simply. Certain plants and certain animals show colour, pattern or form in strong contrast to those of their usual surroundings, giving them, according to the valuations of human sight, a particularly arresting, conspicuous or incongruous appearance. Following the line of thought that leads to a teleological explanation of 'protective resemblance' these spectacular phenomena are usually interpreted as warning signals intended to convey to potential predators the undesirability of molesting their possessors, though the reasons for this are seldom precisely indicated.

It is well to recall here that the three most usual meanings of the word 'warn' are, to give notice of danger; to notify in advance; and to caution. All three rest upon an ability to apprehend the future and to realise that action in the present may spell disaster to come, and this ability is one of the signs of dawning human intelligence and one of the principal features that distinguish man from other animals. The whole behaviour pattern of non-human animals points to the fact that they do not possess this faculty in any reasoning, as opposed to instinctive, way.

The explanation of 'warning colouration' usually suggested is based, in the same way as is 'protective resemblance', on human estimates of perceptual values and the criticism of this need not be repeated, but there are one or two particular points about 'warning colouration' that call for notice. One is that many of the colour and form designs involved are, by more ordinary standards of biological design, bizarre, and it is sometimes necessary in order to apply the familiar teleological explanation of them to use comparisons which are scarcely less bizarre themselves and which bear little relation to plausibility.

Another difficulty is that some 'warning colourations' are almost the same in two or more animals or plants so that any explanation of them must take account also of 'mimicry', a subject sufficiently puzzling in itself.

It may also be pointed out that there is no real evidence, other than that deriving from human perception, that the dangers against which 'warning colouration' is thought to caution actually exist or that prey is avoided by predators because the latter have learnt by experience not to molest the former; and there is no real indication that creatures showing 'warning colouration' have any biological advantage over those of their immediate like which do not show it.

Nor is the idea of 'warning colouration' confined to the matter of sight. Many plants and animals emit, or contain within their bodies, substances which, to human beings, are unpleasant or worse in taste or smell, just as there are others in which the emanation is pleasant or attractive, and it is too often assumed that the former are deterrents, but for this view there is no real evidence, or even probability, at all. Many kinds of insects, especially beetles, are coprophagous and others, including the larvae of certain flies and the 'burying beetles' are carrion feeders, and all these find congenial, and presumably satisfactory, habitats in circumstances repellant to the human senses but it surely cannot be maintained that the occupation of these particular habitats is in any way particularly protective or deterrent. On the other side of the picture the scents of plants, especially those of flowers, range, in terms of human estimation, from the exquisite to the abominable, and even in some cases resemble those associated with carrion-feeding animals, but there is no good reason for interpreting these and other similar facts as having the same values for plants and non-human animals as they have in the aesthetic appreciations of man.

If one conclusion above all others can be drawn from the foregoing discussion it is that there is much evidence to show that the sensory perceptions of animals (and, it may be supposed, of whatever corresponds to these in plants) varies greatly, as is clear enough from what has been said about vision. It has even been suggested, in the concept expressed in the German word 'merkwelt'*, that each and every kind of animal lives in a perceptual world of its own, gaining from this its own particular appreciation of its more immediate environment and of the oppotunities which this presents for exploitation. If this is so then perceptual ability must be reckoned as one of the great biological variables and a whole new field of evolutionary study opens, investigation of which is likely to lead to astonishing results.

To sum up, it is plain that teleological explanations of what are commonly called 'protective resemblance', 'camouflage', 'mimicry' and 'warning colouration' are inadequate, not least because such explanations cannot reasonably be applied to plants, and it becomes necessary to consider what other interpretation can be offered in their place.

It is here that the Angiosperms, and to a lesser but still notable degree other groups of the plant kingdom, go far towards suggesting an answer because a careful and comprehensive study of superficial resemblance in these living things leaves an enduring and deep impression that they are not primarily of a teleological nature but are the consequences of repetitive evolutionary achievement in which a comparatively small number of different basic structural and chromatic patterns are concerned. This repetition suggests that there are, within any area of detailed evolutionary elaboration, certain directions in which evolution has an innate tendency to move. In short they strongly indicate that some sort of orthogenesis has been a primary conditioner of evolution.

This appreciation is plainer if there is recalled the familiar picture of the course of evolution as a much-branched tree that continues to grow but which is always

---

* v. Uexküll: *Umwelt und Innerwelt der Tiere*, Berlin, 1921.

almost totally submerged by a rising flood that represents the passage of time, and in which the uppermost branchlets emerging from the water represent the plants and animals of the present. If this pictorial analogy is acceptable its greatest importance lies in the conclusion to which it points; that proximity among the emergent branchlets is not necessarily a measure of their true genealogical connections and relationships, which remain hidden below the rising water.

The more particular application of this picture to the concept of orthogenesis can best be expressed in a rather different mental image. In this the various directions of evolutionary change can be visualised as a series of railway lines which for the most part run roughly parallel to one another, and each of which has, here and there, one or more branches or sidings. Usually the trains using the lines are 'through' trains but every so often a switch is pulled and some part of a train is 'slipped' in another direction which will, in all likelihood, converge towards one of the other railway lines, so that *superficial resemblance*, as represented by this closer proximity, results. If this switching process is imagined as occurring with varying frequency in various directions in the different lines, then it can only be a matter for wonder that superficial resemblances are not even more numerous than they are. The real problem is what 'pulls the switches' and one answer to this question is most likely to be found among the more refined aspects of genetics.

Part III
The Origin of the Land Biota

The colonization of the land surfaces of the globe by plants and animals of marine ancestry . . . water . . . comparison of life in water and in air . . . transmigration . . . the wracks . . . the retention of the female gamete . . . the tides . . . the thallose liverworts . . . plant plankton . . . the homogeneity of the marine flora in comparison with the marine fauna . . . the relation of structure to life in water and life in air . . . exoskeletons and endoskeletons . . . animal foragers . . . reciprocal transmigration . . . the special problems presented by the fishes . . . the essential differences between the marine and terrestrial environments . . . continuity, uniformity and physical values . . . differences in sensory perception in water and in air . . . some examples of change with time in the terrestrial biological environment . . . tides, continental drift, polar wandering . . . movement in the marine and terrestrial environments . . . the 'carpet' of land vegetation . . . the evolution of flying organisms . . . three particular and illustrative aspects of evolution on land, i.e. frost, flight and grass.

The Origin of the Land Biota

A comparative study of the plants and animals living today and of their physiology, together with what is known about their geological history, leaves little room for doubting that the principal event or sequence of events in their evolutionary history has been the colonization of the land surfaces of the globe by plants and animals of marine ancestry, and this great subject of 'the origin of the land biota' cannot be better introduced than by a short account of the substance which is fundamental to almost every part of it – water.

Water (H_2O) is unique among chemical compounds of simple formula in that it is a liquid at what may be called ordinary, biologically acceptable temperatures and is tasteless, colourless and odourless. Chemically it is neutral in reaction and very stable, and it has unrivalled powers as a solvent, which means that a great many other substances, some very common, dissolve in it. Another peculiarity, in some ways the most far-reaching in importance, is that, unlike almost all other natural liquids, it expands as it freezes so that its solid form of ice floats on its liquid state. Because of this ice rises to the surface of the water from which it comes, and to which it therefore affords some insulation, with the result that it takes prolonged and intense conditions of freezing before any but the slightest depth of water freezes solid. Even in the Arctic Ocean the average thickness of ice is only about eleven feet and the temperature of the depths below it does not fall to freezing point. Its expansion on freezing has the further consequence that materials containing much water, as are most of the rocks forming the immediate surface of the earth's land, tend to break up and disintegrate under the great pressure of this expansion. This last effect is, biologically, of special importance because the bodies of plants and animals are built up of cells most of which are inextensible bags of membrane filled with watery fluid and which, when they cannot expand sufficiently, are liable to burst if this fluid freezes. This is why prolonged exposure to frost is fatal to land plants and animals unless thay have some special means of protection against its effects. All these characteristics of water add up to the fact, and can be expressed in the phrase, that the chemistry of life, which is largely that of the protoplasm which is its physical basis, is very much the chemistry of dilute aqueous solutions.

For all these reasons it is generally accepted that life on earth must have begun in water, which for all practical purposes means the sea, and this basic conclusion is the sheet anchor of studies in evolution. From it follows the opinion that the gradual colonization of the land has been one of the most significant chapters in evolutionary history. This vast subject, of which, because plant life precedes animal life, the most important expression is usually referred to as 'the origin of the land flora', or, more portentously, as 'the subaerial transmigration', was particularly in the forefront of biological thought some fifty or more years ago. It raises many problems that bear profoundly on the nature of the processes of evolution, but before passing on to these it is helpful to remember that there is one fundamental question to be asked. Why should there have been this transmigration and why has evolution not confined itself within the medium where life began; which was for long its only environment; and with which it is so closely in tune? The answer to this question has still to be found, but to pose it is not unprofitable because it highlights the remarkable commentary on the whole process of change with time that it should, after having brought about the exploitation of the aquatic habitat by living creatures, have gone on to what seems, to us at least, a most difficult and hazardous development, the modification of biological form and function so that these are able to express themselves on land and in the air.

It is a common observation that most forms of marine or freshwater life can survive for only a short time when removed from their natural surroundings. This is because the terrestrial and atmospheric environment is not, as is the sea, saturated with water, and, consequently, living bodies insufficiently protected against water loss by evaporation, wither or dry up, with fatal effects, in a comparatively short time. The two environments, marine and subaerial, are therefore in a measure antithetic, and the inhabitants of the one can live in the other only if, in effect, they take part of their more proper surroundings with them.

Human divers are, perhaps, the most familiar example of this because they can live below the surface of the water only if they enclose themselves in a water-tight envelope containing a supply of air; that is to say they take a tiny amount of the atmosphere with them. Again, plants and animals can live in the atmosphere only if they have an airtight covering or membrane so designed that loss by evaporation from their water-saturated bodies can be controlled or prevented. Conversely to the divers they can live on land only by ensuring that their aqueous contents shall not become dangerously desiccated in air. It is the remarkable sequence of structural and functional change in fundamentally aquatic bodies, making them more and more resistant to drying up, which is the essential theme in the origin of a land flora and, by extension, of the land biota.

Some idea of how the actual process of 'transmigration' occurred can be gained from a brief account of certain of the larger marine algae, or seaweeds, of today, and particularly of those popularly called 'wracks' (members of the genus *Fucus* belonging to the Brown Algae or Phaeophyceae). Except when exposed by the ebbing of the tides these plants, like most other seaweeds, live immersed in sea-water, from which they obtain the natural supplies they need. Consistent with this their structure is simple, the plant body being little more than a branched and flattened sheet of tissue, commonly provided with pockets of air to give buoyancy, attached to some solid substratum by a simple adhesive holdfast. There is no root system because the plants, which have no impervious cuticles, are normally surrounded by water, and by the same token there is virtually no conducting system for carrying substances from one part of the body to another. Yet the dimensions of these plants are to be measured in feet, and some of their most similar associates reach far greater sizes. Their reproduction is likewise very simple, eggs and sperms being produced in enormous numbers and exuded from the parent body directly into the surrounding water, where fertilization takes place. The 'wracks' thus perform their essential functions of growing and perpetuating their kind with what, considering their size must be regarded as almost the simplest possible apparatus, while their great abundance on the shores of much of the temperate regions, together with their obvious ability to thrive in the rough conditions of tidal waters there, show that their whole structure and mode of life is geared to their environment to a quite remarkable degree. In their combined simplicity of form and reproduction they may surely claim to be the most efficient life machines in existence, their only rivals for that title being other kinds of seaweeds with similar life-histories.

Unless land plants are to be regarded as something *sui generis* and as having come into existence independently, it must be accepted that they have evolved from marine plants, and that the seaweeds and some of the least differentiated land plants present an almost unbroken series of forms, marked by an increasing ability to withstand exposure to air, or to be more precise, an increasing ability to dispense with surroundings of liquid water during parts of their life-cycles, is good evidence of this. This is very simply illustrated in the zonation of the seaweeds themselves on tidal shores, where the various sorts occupy different levels within the tidal range according to their ability to endure periodic and inconstant exposure to the atmosphere, those most vulnerable to this being attached near or below the lowest levels of spring tides; those least vulnerable at levels near the upper levels of spring tides; and the rest appropriately in between. Such zonation certainly gives good ground for believing that the actual migration from water to land was achieved by the gradual evolution of algal forms more and more tolerant to exposure until at last there appeared plants that were entirely free of any dependence on aquatic immersion.

This account of the 'wracks' makes it clear that one essential feature in their story is the existence of marine tides, as a consequence of which the sea-level at any one place on the shore-line rises and falls periodically, usually twice in twenty-four hours. This has two main practical results the significance of which in biological affairs can hardly be exaggerated. It means that parts of the sea-shores are regularly, but in varying degree, exposed to the air when the tide falls; and that there is a periodic mass movement of water which, because of the intercommunication between the sea areas of the globe, spreads far and wide, with a general distributory effect. The rotation of the earth alone would be enough to cause some marine tide but that the tides are as variable and yet as precise as they are is due mainly to the fact that the earth has a satellite, the moon, and it is the relative conjunction of the sun and the moon which gives the tides their peculiar, and biologically invaluable, rhythm, in which the amplitude of the rise and fall waxes and wanes over a given period. The tides also have many more minor consequences and it is safe to believe that they have exerted a profound influence on the processes and course of evolution.

Before taking leave of the wracks there is another point about them to be noted, the particularly simple way in which they illustrate difficulties in any idea of 'natural selection' in its classic form. There is no reason to suppose that the populations of shore-attached seaweeds vary much from generation to generation, and if this is the case it follows that, of the vast numbers of propagules these plants produce, almost all perish. Moreover those that will survive have first to be satisfactorily fertilized and then attached to a suitable substratum. Only after this can they begin to grow into the plants which will have to withstand the storms and stresses of their turbulent environment before they themselves can mature and reproduce. That some succeed in doing this is self-evident since the populations remain more or less constant, but that these 'surviving' individuals are somehow superior in characteristics to all the rest, or that the propagules from which they come possessed some advantages over the myriads that perish, is not credible. It can of course be argued that the mere fact that few survive while many perish is a process of sifting which can, by an imaginative stretch of definition, be called selection; it may also be argued with some justification that a sifting like this does in fact help towards steering the course of evolution in certain directions, but this is indeed a far cry from the idea behind Darwinian 'natural selection'.

To return to transmigration, the next stage in the movement into the atmosphere is generally thought to be represented today by some of the least differentiated plants of the land, the thallose liverworts, which have bodies not very different from those of the wracks but which, unsupported by surrounding water, live flattened against a substrate and attached to it by absorbent hairs more or less throughout their lengths. More significant, however, is the fact that the female gametes are not shed from the parent plants but are retained within their tissues, where they have greater protection from desiccation. This retention of the female gamete is one of the prime themes in the development of plant life on land, and it is therefore of more than ordinary interest that a similar kind of retention is to be seen, as the 'carpogonium', in some of the most seaweedy of seaweeds, the more delicate red algae, suggesting that the first land plants developed an inherent tendency which had made itself apparent elsewhere long before, but which had not previously been fully exploited.

Despite the importance that the larger seaweeds may have had in the process of transmigration it seems reasonable to believe that, in the earliest times, plant life in the sea consisted of very simple floating organisms able to nourish themselves autotrophically by virtue of the presence in their bodies of chlorophyll-like catalysts (see p. 123). If this is so then their scope of life must have been restricted in ways which ordained much of their evolutionary future. Photosynthesis, as its name implies, requires light, and for this reason the first plants must have been capable of floating in the upper layers, and perhaps only the uppermost few fathoms, of the water, through which sufficient light could penetrate. They must, in short, have been what today we should call 'plankton'. Later on, and probably much later on, some of these floating forms became sedentary upon the only substrate available to them, the tidal shore zones. Here, unhampered by the requirements of free flotation, they gradually developed into larger and more differentiated plants, of which the wracks are some of the most noteworthy. The coastal zones are, compared with the vast surfaces of the open ocean, almost negligible in area, with the rather paradoxical result that the marine plant life most obvious to human eyes is of relatively small account in the total biological economy of the sea. Noting, as we so easily can, the quantities of large seaweeds growing, or cast up, on our shores it is clear that these provide a potential source of much animal food, and much evolution in marine animals must have been based on it, but the free-floating planktonic world which, it seems likely, existed ages before any attached seaweeds had evolved, must have been, as indeed it remains, of much greater significance.

Today plant-life in the open sea still consists almost entirely of algae, either free-floating or anchored where circumstances allow. Beyond this stage the evolution of truly marine plants has not progressed, and we are hardly likely to be wrong in thinking that one reason for this is that plants of algal organization, such as have been described, fulfil their functions in the sea so efficiently and completely that they cannot, to use common parlance, be improved upon. When in the course of time, however, colonization of the sea shores gradually became intensified and the problems of exposure to air became more pressing, this no longer sufficed, and a great new chapter opened.

At this point it is interesting and instructive to notice the remarkable homogeneity of the marine *flora* compared with the marine *fauna*. Most outline classifications of the animal kingdom divide it into ten or a dozen major sections, or *phyla* as they are called, each equivalent in terms of classification to the whole of the Algae among plants, and of these phyla all are represented in the sea, one or two of them almost or quite exclusively; others in a closer balance between sea and land; and in only two of them are the numbers of forms notably greater on land than in the sea. These are the Arthropoda, among which the land insects greatly outnumber the marine Crustacea; and the Chordata, of which all but a few are vertebrates. Among the latter, which comprise fishes, amphibia, reptiles, birds and mammals, the range of form and mode of life is much greater on land than in the sea, but as far as can be assessed from man-made classification, the total number of kinds of land vertebrates does not exceed that of the fishes of the sea.

One of the chief requirements of life in the sea is some power of flotation or buoyancy which will allow the greatest possible amount of body surface to be exposed to the surrounding water and, in the case of plants, to light, but on land the comparable need is for some structural design which will allow the body, be it plant or animal, to remain extended in the atmosphere. This is, obvious enough in trees and other plants which have rigid and divaricate stems, and there can be little doubt that the predominance of the Arthropoda and Vertebrata on land is linked with the fact that the members of both these groups possess analogous structural support. In the former this takes the shape of a hard, resistant, dermal covering, or *exoskeleton*; and in the latter of a bony internal framework or *endoskeleton*.

It is noteworthy that although the Arthropoda, with their exoskeletons, outnumber in assemblages all other great groups of living things, terrestrial or marine, they have never achieved any great individual size; indeed some existing marine crustaceans are larger than any insects or other members of the phylum now living on land. That this should be so no doubt involves profound problems of physiology, but it is also clearly related to the fact that an animal with a complicated exoskeleton can increase appreciably in size only at such times as this covering is soft and extensible. This, in practice, involves the animals in 'casting' or 'sloughing' their exoskeletons at intervals during their lives and in growth being restricted to the periods after these castings and before the new exoskeleton has had time to harden. This routine is seen in simple outline in the marine crustaceans, but in the insects of the land it has become developed into a most elegant series of variations on the theme of metamorphosis in which the whole life-cycle is divided into stages, most familiar as larva (grub), pupa (chrysalis) and imago (perfect adult). In contrast the typical endoskeletal vertebrates are under no such handicap since their supporting framework allows them to grow, if necessary continuously, in all three dimensions. Hence, many land vertebrates attain considerable size, but this cannot be carried beyond a certain point because great weight and the necessity of supporting it in the atmosphere ultimately impedes the capacity for movement, and it is left to some of the marine mammals, the whales, in which great weight is supported by water, to claim the prize for sheer size and bulk. It is in accordance with this that some of the largest extinct reptiles are commonly supposed to have been inhabitants of swamps rather than dry land proper.

In this connection it is of great interest that a minority

of vertebrates, notably the turtles, tortoises, and the pangolins and armadillos among mammals have, by the specialization of their endoskeletons, achieved an exoskeleton also, but adjust growth to it rather differently. In some fishes, too, the normal scaly covering is more than usually protective, but here the principle involved, especially in relation to growth, is somewhat different.

Earlier pages dwelt on the *likenesses* between plants and animals because these give good reason for believing that evolution has, in general terms, been the same process in both, but there are also certain basic *dissimilarities* between them which, though they do not outweigh the similarities, throw much light on the way evolution has expressed itself differently in the two and which bear directly on the question of transmigration. Most significant here are the differences in the methods of nutrition in plants and animals.

The word passive is rightly applied to plants because their food supplies come to them by solution and gaseous diffusion, and they need therefore take no active steps to obtain them. Animals, however, depend directly on plants for their food or indirectly on other animals which are themselves plant-consumers. The virtual immobility of plants therefore leads to a situation in which animals must necessarily be able to move in order to search for their food, because an immobile animal would soon exhaust any supplies within its reach. So it is that animals are, in general, *foragers*, and this involves much structural variety and elaboration. For example marine and freshwater animals must be able to swim or exhibit movement in some other way even if this amounts to no more than drifting or the production of local currents in the water; and land animals must be able to move readily on, or just below, the surface of the ground or fly in the air. Indeed it is fair to say that the evolution of animals has, in large measure, been a matter of the development of the powers of movement and locomotion.

The last few pages have, it is hoped, given some

impression of the 'subaerial transmigration' as one of the great episodes of evolution. It is a conventional outline of the subject based on the premise that life began in the sea and underwent its earlier evolution there, but there is no reason to doubt that it substantially records the actual course of events. It is, however, only an outline, and several of its aspects merit further comment, outstanding among these being the question of how far there has been a *reciprocal* transmigration of land organisms into the sea.

That some of the animals living wholly or partly in the sea are the descendants of others which were evolved on land is generally accepted and is exemplified especially by the whales and dolphins, of which many kinds are entirely pelagic. Other, less extreme, examples are the seals and walruses which pass most of their life on shore and can be described as littoral. These are all mammals but much the same kind of life is led by the turtles, the sea-snakes, and one or two other reptiles. Among birds the penguins show a particular form of this life-style which makes them some of the most specialized of vertebrates. The amphibia for the most part reverse the pattern, living on land and reproducing in water. This leaves, among the vertebrates, the fishes, and here there are kinds that occur both in sea and freshwater. Among the invertebrates there are examples of most of these life-patterns but the most prominent and notable fact here is that the insects, so predominant on land, are virtually absent from the open sea.

The question of whether there has been reciprocal transmigration from the land into the sea, and if so how widespread it may have been, is best approached by referring to one particular phase in the history of the land environments of the world as described in the outline of the 'soup theory' given later in this book. This theory postulates that the ocean waters were relatively pure until such time as precipitation of water from the atmosphere, predominantly in the form of rain, carried products of land erosion into the sea and so initiated a new and much accelerated stage in its salinification. What is inherent in these accounts of this period but is not always made sufficiently plain is that this erosion must have resulted in the gradual development of the world's land drainage systems: it produced the great river network of the land masses. While the transition from sea to subaerial land must generally speaking be abrupt, that from the sea into the freshwater of rivers is something very much more gradual and, most important, does not involve a fundamental change in the fashion of respiration. In other words the river systems provide lines of infiltration penetrating deeply into the land masses. The further details of the 'soup theory' make it clear that this great network of freshwater first took form an incalculable time before any of the great groups of plants and animals were evolved and was therefore ready for exploitation by them when the appropriate time came. In consequence when the various great groups of plants and animals began their slow move towards the land they had, in the river systems, lines of approach that they could follow deep into the continents without serious obligation to evolve new methods of respiration and new forms of locomotion. It would therefore seem a reasonable inference that, long before any truly 'land' animals appeared, aquatic animals may have made their way deep into the land masses. What and which these animals were is not precisely known, but it is clear enough that those with considerable powers of locomotion would find their progress greatly facilitated. Other groups, depending on less efficient methods of movement (including the algae on the plant side) would have been much less able to take advantage of the conditions.

An important point here is that rivers not only flow comparatively swiftly but, near their mouths, are not only influenced to some degree by the tides but frequently provide areas in which there is high variation in such environmental factors as water temperature, salinity, water flow and water depth, all of which contribute to conditions generally regarded as likely to be particularly favourable for rapid evolutionary change.

Penetration inland by rivers would be most easily achieved by animals capable of moving against the current by muscular action, and two great groups in the marine biota, the crustaceans and the fishes, particularly meet this requirement. It is tempting to pursue the case of the former and to suppose that the migration of crustaceans into freshwater marked the birth of the great land animal group of the insects, but it is the fishes that raise special problems in this connection.

The fishes today make up the only vertebrate group that is predominantly marine and, with the exceptions mentioned earlier, they are the only vertebrates of the sea. The broad facts of this situation are so well known that it is easy to under-estimate the anomaly that it presents. On land and in freshwater all five great vertebrate groups are well represented (though the birds are less aquatic than the others); in the sea there is virtually only one of them, but this group is generally thought greatly to outnumber its land and freshwater counterparts in numbers and variety.

Appreciation of this situation is the more difficult because it is generally believed that the fishes are the most primitive vertebrates and the group from which the others must have evolved, but this view rests in part on the belief inherent in the concept of the subaerial transmigration that the marine fauna has given rise to the fauna of the land. This consideration, and many other varied morphological and physiological facts almost inevitably raise the question of whether the fishes, *in toto*, are the primitive marine vertebrates that they are often held to be, and whether the marine fish fauna may not have been contributed to more largely by reciprocal transmigration from the freshwaters of the land masses. Such an idea can be pursued in several directions and it has even been suggested that the fishes of the sea may have been evolved from fishes of freshwater rather than the opposite.

Many objections would have to be overcome before this suggestion received wide support but it is important because, if substantiated, it would not only greatly alter but also much simplify the general picture of animal evolution by allowing that all five great vertebrate groups evolved on land, thus removing the anomaly of a huge vertebrate population of one class in an otherwise almost wholly invertebrate biota. It would also help to explain why so many animals with an endoskeletal design, appropriate to animals living in the atmosphere, should be found in the sea.

How the sea has come to possess its present piscine population is puzzling not least because the fossil record of these animals is unusually long, dating from very early stratified rocks, and copious, and some of the earliest fish types are known chiefly from their external features only, which makes it difficult to establish their relationships. There are, however, some conclusions that appear to be indisputable and these afford indications of great interest. First to be mentioned is the existence today of a small group of animals called lancelets and belonging mainly to the genus *Amphioxus*. They appear to be the simplest fish-like animals now extant and they are marine, but live in shallower waters. The first 'real' fishes are recorded as whole or incomplete fossils and are classified as the jawless fishes because they have no jaws or teeth of the ordinary kind. Most of them have long been extinct but the Cyclostomata, which include the lampreys and hagfishes of today and which are partly marine and partly fresh-water, are held to be their present representatives. These animals are soft-bodied and scaleless; have no paired fins; are without true bones, though their bodies are somewhat reinforced by cartilage; and they have one or two other peculiarities.

The present-day fish fauna comprises two main sorts of animal, the selachians and the rest. The former (sharks, dogfishes and rays) are known from many geological horizons from the Devonian onwards, and have cartilaginous skeletons without true bone. They are virtually wholly marine and are characteristically pelagic. With very few exceptions provided by the persistence of a small number of geologically much older types, the remaining fishes of today are termed in general teleosts or bony fishes because they have well-developed, and often all too obvious, bony skeletons. These have been classified in great detail and show a remarkable range of form and life-pattern and included in them or at least closely related to them is one particular small group of three genera of lungfishes, which live in freshwater but which are able on emergency to survive for a time in air. Except for the Cyclostomata all freshwater fishes belong to the bony fishes and this great group provides nearly all the range of form of the fishes in the sea. The bony fishes are geologically younger than other kinds, their first fossils being recorded from Mesozoic rocks.

Just what sequence of historical events these data reflect is hard to determine, but it seems reasonable to conclude from them that the first 'fish' were marine animals without jaws and without bone; that these were, in course of time, followed by more distinctly fish-like creatures, classified today as Placoderms because of their armour-like external covering, some of which lived in freshwater, but that these had all become extinct by the end of the Palaeozoic; that these were succeeded by selachians which attained their now-familiar form in the Mesozoic; that the true bony fishes appeared also in the Mesozoic, by the end of which time they had become the predominant marine fish; and that these bony fishes are today, except for the Cyclostomata, the only sort of fishes that live in freshwater. None of these points exclude the possibility that the bony fishes of the sea have evolved from freshwater fishes and if the time sequence of the many fossils of bony fishes could be firmly established this question, which is crucial to the whole problem of the evolution of the vertebrates, could no doubt be solved.

It is not the primary purpose of these pages to discuss the hypothetical history of the fishes but the subject throws useful light on so many wider aspects of evolution that one final comment on the subject is justifiable. If all the possibilities are to be considered one must certainly be that an ancestral form of what became the fishes, which is now most closely represented by *Amphioxus*, evolved in the particular environment presented by the shallow, and often tidal, waters of the continental shelves; that these ancestral forms gradually developed along two distinct lines, one leading to the selachians, which proceeded to exploit the deeper waters of the oceans proper, and the other leading to the exploitation of riverine waters by what became in due course the bony fishes; and that some of these latter, being well designed to do so, later also exploited the oceanic waters in a reciprocal or reverse transmigration.

This discussion about reciprocal transmigration affords a good opportunity for reviewing the contrasting features of the marine and subaerial environments, some of which have already been touched upon. These comparisons can usefully be summed up under the three headings of continuity, uniformity, and the physical differences between water and air.

The first of these is the most obvious because the land masses of the earth, whether large or small, are all in fact islands in that they are surrounded by water, and hence this circumambient environment is continuous, every part of it communicating with every other. This continuity has however some limitations chiefly because the depths of the sea varies greatly, and many of these communications are by shallow water only, and also because the relief of the sea bed is pronounced, a point mostly expressed in the great ridges which traverse many of the ocean basins, and the many deep water trenches which occur here and there. Nevertheless continuity, at least in comparison with the land, is one of the chief characteristics of the marine world.

Linking continuity with uniformity is the fact that the temperature of the sea varies but very gradually from place to place and the equator-poles gradient of temperature is much less steep than it is on land. Moreover the sea, except for some surface waters at high latitudes, never freezes and its inhabitants do not therefore have to face the problem of frost.

Uniformity of environment in the sea is largely attributable to its continuity, and it is the lack of such continuity on land which gives the latter its most characteristic biological feature, *diversity of habitat*. The most striking expression of uniformity in the sea is its salinity: everywhere the waters of the oceans contain a variety of dissolved salts which give it a salinity of about 35 parts per thousand, and only in seas closely associated with adjacent land is this figure greatly exceeded or diminished. The salts of sea water are derived from twenty or more elements of which chlorine and sodium greatly outweigh the rest, followed by magnesium, calcium, sulphur and phosphorus in that order.

The physical differences between the marine and subaerial environments are more difficult to describe and can be mentioned here only briefly and in so far as they illuminate the problems of subaerial transmigration, from which point of view they express themselves most cogently in terms of density and pressure. Of particular note here is the phenomenon of osmosis which, in both plants and animals, but more directly in the former, is a matter of prime biological moment. Most marine invertebrates are isotonic, the osmotic values of their internal fluids more or less balancing those of the waters around them, but creatures of the land cannot, by the nature of things, be so, and the control of osmosis and its effects, and the development of the means by which this is done, is one of the leading themes of the subaerial transmigration.

For the rest no more can usefully be done here beyond considering, in very general terms, the effects of the differences in density and pressure in the two environments on the senses of the organisms concerned.

To take them in their most convenient order the five senses commonly recognised are tasting, hearing, feeling, smelling and seeing; but first of all it must be plainly stated that the actions and reactions of animals in general (and whatever their parallels in plants may be) cannot be satisfactorily explained fully in terms of these five only and the following short survey must, on this count, be considered incomplete. Just what other and additional senses there may be among non-human animals remains to be discovered, and doing this is likely to be difficult because man can work only from the basis of the senses which he himself experiences.

Tasting would seem to be the sense least likely to be affected by the change from sea to land, except that, in accordance with the vastly greater range of environmental conditions on land the gamut of flavour is probably much wider there than in the sea. This suggests the interesting question of how far what man regards as poisonous plants and animals (others than those of the latter in which poisonous stings or fangs are defensive or offensive against other animals) make their presence felt in feral nature. To put the point rather differently to what extent does accidental poisoning from the consumption of toxic plants and animals occur among truly wild members of the biota? Certainly in temperate regions, where the facts are most easily observed, the environmental distribution of many plants commonly held to be dangerously poisonous to man suggests that it is not frequent, a conclusion that relates to some of the problems of superficial resemblance discussed earlier. Also to be borne in mind is the question of whether the actual mechanism of taste is the same in all animals.

Hearing is particularly difficult to assess because of the extent to which it involves highly specialized organs, some of which are different in structure from, but appear to perform the same functions as, those to which the word 'ear' commonly applies. The fish, for example, and to a lesser degree the amphibia, lack much of the complicated ear structure characteristic of birds and mammals. Nor is the problem simplified by the various mechanical methods of transmitting sound through water which are now an important feature of navigation, because there is no good evidence that these reflect any actual biological situation.

It is also in this connection that there arises the question of how far what the human ear detects as 'noise' is a natural feature of submerged marine life, and whether creatures in that medium communicate by sound in anything like the way they do on land.

Feeling is in much the same case but here there are two particular points to remember. Feeling in water is to some extent the effect of vibrations transmitted through the fluid and would seem to have a close affinity with the sense of hearing, though something akin to this operates on land under such influence as wind pressure; also the actual perceptivity of feeling is likely to be more delicate in the air than in the water.

Smelling seems to present a rather simpler picture, assuming that this sense is possessed in ordinary fashion by all marine animals, because the sea can hardly be as receptive to the long-range wafting of scents as is the air. In fact there is good reason to believe that perceptivity of smell has been greatly developed in many land animals, as is evident from man's experience of the animals he has domesticated and the quarries he has long hunted. Indeed it may be argued with some confidence that smell is the sense in which man has most clearly fallen behind in the course of his development.

Finally there is sight and there can be no doubt that it is this sense that has been most affected by the change from a marine to a subaerial environment, if only because in the latter the necessity of, and the potentiality for, long-range vision make themselves particularly apparent, and this impinges on almost every aspect of post-transmigrational evolution, and again not least on the problems of superficial resemblance.

As to the question of additional senses the most promising subjects for its investigation are almost certainly the insects, those predominantly 'creatures of the land'. That their evolution has been largely, if not wholly, a consequence of the subaerial transmigration can scarcely be gainsaid, and this being so their unrivalled variety and ubiquity can be regarded as one of the most prominent of all the results of that great change, and the most likely to have involved the evolution of senses other than those commonly recognised.

As has already been explained, the basic characteristic of the marine environment is its uniformity in space and chemical constitution, as well as in time, for it changes but very slowly. Sea-water is virtually the same in almost all parts of the ocean, the most notable differences being those caused by the superficial effects of the climatic gradient between the equator and the poles. On land exposed to the atmosphere, and in the atmosphere itself, circumstances are almost completely different and, in particular, there are, in the subaerial environment, two major gamuts of value scarcely apparent in the sea. One is the amount of water, as vapour or liquid, in the air and in the soil; the other is the physical and chemical make-up of the soils themselves in which plants of the land must find anchorage and from which they must draw many of their requirements. Hence invaders of the land from the sea encounter and must, if they are to succeed, come to terms with, two series of unfamiliar circumstances, one consisting of differences in humidity and temperature, the other of differences in the physical and chemical conditions of the substrate. Together these present a range of influence, scope and opportunity much greater than anything that the sea has to offer. They are the chief contributors to the great heterogenity of the land environment and this heterogeneity cannot fail to have increased, not only the scope of evolution, but also its speed, because of the enormous range of new habitats that it offered.

Apart from these more familiar factors three others, inherent in the form and nature of the earth as a heavenly body, are particularly worthy of note. One is the phenomenon of the marine tides, some of the consequences of which have already been discussed on p. 85. Another is continental drift, the differential sliding or creeping movement of various of the land masses over the deeper material of the surface of the globe. The third is the wandering of the poles, which can most simply be described as the long-term 'wobbling' of the axis of the rotation of the earth. Biologically the most important consequence of this is that the equatorial line, which is the edge of the plane at right angle to the axis of rotation and which is at all points equidistant from the poles, moves over the surface of the globe according to the amplitude of the polar movement. Should this figure reach 90 degrees of latitude the equator would lie at right angles to its present position and through what are now the poles. If the surface of the earth was completely homogeneous and undifferentiated the effect of this might not, bearing in mind the immense length of the time factor involved, be very great, but since it is divided up, mostly irregularly, into areas of sea and land, the latter varying greatly in relief, the proportions of land and sea across which the equator would run is likely to alter much with the passage of time, and this is likely to have influenced the development of the earth's biota in many ways but especially through the gradual movement of the major climatic zones. Of these three phenomena the two latter would in combination be capable of producing an enormous range of heterogeneity in the land environment, including no doubt some conditions which have not as yet been expressed.

It is not difficult to picture the gradual multiplication and elaboration of form with time among marine plants and animals which have always lived in their aboriginal, and comparatively simple, surroundings; and the process itself may be regarded as comparatively simple in that it is unlikely to have involved, of necessity, any sudden departures from established principle and practice but rather to have consisted of the continued exploitation of trends initiated very early. It may even be felt that given time and the cumulative effect of variation, multifariousness in the marine biota was a foregone conclusion, and that even the first tottering steps towards the land, the brief and occasional exposure to the atmosphere in the inter-tidal zone, were only an extension of this, but when totally subaerial life is considered – life passed from start to finish entirely dissociated from direct marine contacts – it is clear that the first transmigrants must, in the new environmental heterogeneity, have been faced with a whole series of problems resolvable only by the development of many new abilities and structures, and how, and by what agencies, these can have been achieved is one of the basic problems of evolution.

Nowhere can the change from sea to air have led to more immediate, and far-reaching, consequences than in the sphere of movement. The earliest transmigrants, having originated in the sea, found themselves the possessors of a vast and novel estate, and merely to occupy and exploit its potentialities, even to the limited extent of maintaining themselves in it, called for the reassessment of at least one fundamental requisite, the power of movement and locomotion.

Movement in the sea can be summed up as of four sorts; the production of local water currents, in which the water is made to circulate round the organisms concerned; drifting, in which organisms move at the dictates of water currents; crawling, on the substrate or on the bodies of the larger seaweeds, a mode of progression chiefly expressed in the shallow waters of continental shelves, and particularly in the tidal zones that fringe the land; and swimming, by which there is active penetration of the continuum by means of muscular action. On land things are very different. Drifting is limited to the caprices of the wind; swimming is, on dry land, precluded; and the only feasible method of penetrating the atmosphere must, at least at first, have been by crawling, incidentally a process without which the actual movement in the transmigration can hardly have come about. But compared with swimming in the sea crawling is a relatively ineffective method of penetrating the subaerial environment, especially since, as seems to have happened, many land animals in due course gave rise to smaller descendants while much of the vegetation became more and more composed of large and bulky plants.

At this point it is helpful to recall that the vegetation of the oceans is wholly thallophytic and is distributed in a simple and distinctive way. Over the vast expanse of the open sea it consists of phyto-plankton which, by and large, floats within the depths to which light can penetrate effectively and this contributes much the larger proportion of the plant life in the sea. Along the tidal coasts of land masses and islands there is a comparatively tenuous and incomplete belt of anchored, larger, thallophytes, the seaweeds proper, which because they are so noticeable to the casual beholder appear to be of much greater importance to the whole marine economy than is actually the case.

On land the vegetation is made up of all the great plant groups but is overwhelmingly composed of vascular plants (ferns, gymnosperms and angiosperms). All these groups together cover such parts of the earth's surface as are habitable with what has aptly been described as a 'carpet' of vegetation. This is a very true description because the plants concerned form a 'pile' of stems and leaves rising from the soil to varying height. In arid regions the pile is short and threadbare; in forests it is closest and deepest, and between these two extremes there is almost every other value.

The very difference in height has its effect, sometimes beneficial and sometimes not, on the animal life of the land and this is enhanced by the fact that, more often than not, the reproductive spores, cones and flowers of the larger plants are borne either towards the apices of the stems or widely distributed over their bodies. Hence they are often some distance above the ground and animals trying to visit or obtain them must cover considerable vertical distances in one way or another.

Many animals, especially the smaller ones that are not so inhibited by their own weights, are capable climbers, but that they are small means that the distances they must cover are relatively greater. There are of course many special cases within this generalization but it is clear enough that the full exploitation of the land vegetation by animals restricted to surface movement only would be a very lengthy process and might well prove to be 'uneconomic' in terms of time as expressed in life-span. One must be cautious in using such a word as 'uneconomic' because it is based on the human concept of 'profit' and 'loss', but that there is in natural evolution a principle which directs the process towards the most appropriate employment of available energy and resources seems indisputable. At least we may believe that there is something of this kind which maintains the energy equilibrium of nature and that results in a state of 'order' very different from the 'disorder' so common in many spheres of human activity.

The problems presented by the air space *within* and *above* the vegetation leads directly to the consideration of one of the most remarkable of all steps in biological evolution, namely the 'invention', for there is really no more appropriate word, of flight. It is not difficult to detect the stages of the process by which certain animals have achieved mastery of the air but it is harder to understand the principles which have ensured this success and it seems to illustrate particularly well the general belief that an urge towards what may be called the 'exploitation of the environment' has been one of the most important factors controlling the course of evolution, and the morphological aspects of this evolution largely reveal how this has proceeded.

Almost any aspect of life on land could be used to illustrate this general point and to give too many examples of it would not only be wearisome but would obscure the narrative, and it must suffice to deal in some detail with three particular aspects of the matter which go far towards covering its whole scope. These three may be summarized in the three familiar words – frost, flight and grass – and may be dealt with in that order.

Partly because of the widely held view that the earth has been cooling for much of its history, and partly because of the peculiar qualities of water, it is generally believed that the climatic conditions in which the first land organisms lived and long continued, were those of much rainfall and of temperatures so high that the inherent dangers of frost were negligible. What those values may have been is not exactly known, but there is good reason to suppose that those of the present equatorial zone, which include an absence of marked seasonal change, are most like them. If so then an important later chapter in the evolution of the land biota must have been the gradual colonization of those parts of the world's surface where conditions were more rigorous, until today the only regions without life in some form or other are where polar conditions of ice and snow prevail or where, on a smaller scale, there is virtually no precipitation from the atmosphere at all.

Just as the earliest land plants and animals could surmount the hazards of desiccation in their new environment only by developing a protective system against undue evaporation, so the dangers involved in spreading into the less hospitable parts of the land, which, in a word, are those where frost occurs, could only be met by the evolution of special structures and modes of life designed to lessen this peril. In both kingdoms of the living world this has been done, to varying extent, by the elaboration of protective outer layers such as unusually thick cuticles or woolly coverings in plants and by winter fur and plumage in animals, but much more widespread in both kingdoms is the practice of hibernation, a mode of life in which, by various means, the dangerous part of the year, winter at high latitudes, can be passed in a state of suspended animation in some protective state or situation. This is most familiar in some animals such as the dormouse; in the cold-blooded vertebrates, and in many insects, but it is much more plainly to be observed in those higher plants which pass the unfavourable months either in the resistant form of seed or in some expression of defoliation, in which the vulnerable leafy parts are shed periodically. In large woody plants this generally involves only the shedding of the leaves, the rest of the plant being protected by the bark, but in herbaceous plants it often consists of the retreat of life from all aerial parts into more or less modified or protected organs below the surface of the ground.

It is a generally held opinion that glacial climatic conditions, in which large ice-caps formed and long persisted at the poles, have occurred but very infrequently during the geological past and that, in the long intervals of time between them, the climate of the world was not only more genial but more equable. The most recent of these frigid visitations – the Pleistocene Ice Ages – belongs to the very latest chapter of geological history and occurred long after the appearance of all the now-existing major groups of plants and animals. If it can be assumed from this that the present flora and fauna of the coldest parts of the world (the Arctic-Alpine) has the shortest geological history and are among the newest of their respective kinds, it may well be that it is among these that some of the processes of organic evolution will be found to be most clearly expressed. Finally with regard to the problem of frost it is interesting to notice that the structural modifications which its solution involves are comparatively minor. This is because hibernation and similar processes are abdications rather than changes of function and therefore do not call for so much alteration.

One of the ways in which the land differs from the sea in marked respect is that it consists of two different but closely associated parts, the rock and soil surfaces themselves, with that diversity of condition already mentioned, and the air above them. Generally speaking plants, and consequently animals, occupy only the shallow uppermost layers of the soil, except that the larger the plants the further they grow up into the air and the more their roots penetrate the substratum. Apart from this it may be said that the air, as such, and other than that part of it actually in contact with the soil and its inhabitants, is scarcely occupied at all. It is true that some small organisms whose weight is negligible are so much inhabitants of the air, at least during their adult lives, as to make up what has been misleadingly called a kind of 'aerial plankton' but this is an over-picturesque interpretation of the situation. As far as is certainly known, no living things pass the whole of their lives, including their all-important reproductive stages, completely independent of and out of contact with the earth or the surface of the sea, or with other living things inhabiting these.

There can be little doubt that the principal reason for this is that plants and animals, having originated in water, lack the innate or inherent potentiality which would enable them to colonise the air in the same sort of way in which they occupied the land, and that to accomplish this they have been obliged to develop, *ab initio*, features hitherto unrepresented in them. At all events the evolution of living things capable of flight seems to have begun much later in geological time, long after the initial subaerial transmigration, and has since progressed very unevenly.

Only three groups of organisms living today, all animal, namely the birds, the bats and some of the insects, have so far been able to conquer the air to the point of making it an everyday part, though only part, of their natural environment, and even in these flight is subject to the seriously hampering limitation that these animals can leave their terrestrial habitats only for such short periods as their energy systems permit. They cannot remain in the air indefinitely because it does not, except in those compara-tively few instances where fliers prey on fliers, afford them any energy-producing food and they are therefore obliged to return to earth frequently for what, in modern terms, is called 'refuelling'. Nor can they rear their young in the air, though some few may mate there.

It is true that a small number of vertebrate animals are able, generally by the development of extensible folds in the skin, to lengthen what would otherwise be simply jumping or springing into glissades, which in the more extreme examples may carry them, volplaning, for a quite considerable distance, but this is akin to what is now called gliding and is not the true flight in which height is maintained or increased by muscular or mechanical power. Nevertheless it is of considerable interest to the evolutionist because the animals which exhibit it, though few in number, belong to three different groups of mammals; one of reptiles; and one of amphibia, showing that a similar trend of development has become initiated in at least five evolutionary lineages. Something of the sort is known also to have appeared in the now-extinct reptilian group of the pterodactyls which, from what one reads, seem likely to have lain functionally somewhere between the gliders and the true fliers. Today's 'flying fish' scarcely merit comparison because their 'flight' is no more than a leaping out of the water and skimming above the surface until they plunge again, nor does their leaping fulfil any comparable purpose of moving from one variety of environment to another.

The larger land plants have succeeded in occupying at least some parts of the lower levels of the air by developing stout woody stems furnished with more or less divaricate branches, and this is relevant here because it puts into its more proper perspective the subject of animal flight, showing that it is essentially an extension of the powers of locomotion enabling the organisms concerned to exploit, more rapidly than would otherwise be possible the vast extent of plant life which rises in varying degree above the surface of the ground.

As has already been written, the vegetation of the land is well described as a carpet in which the depth and closeness of the pile varies much from place to place. In the more arid parts of the world the carpet is threadbare and its depth is almost or quite negligible: in the most favourable climatic conditions its pile is dense and may be more than a hundred feet deep; and in between these two extremes there is almost every intermediate state. Many land animals are accomplished climbers but even so the exploitation of this carpet would, if they had to rely solely on surface movement, be a wasteful procedure from an energy point of view, for it must be remembered that land animals have nothing comparable with the air-pockets which help to keep the larger seaweeds fully extended; neither are they balloonists, either tethered or free. Surely therefore it is a most remarkable aspect of evolution that many land animals have developed an even more astonishing means of speedily crossing intervening spaces in the atmosphere, namely muscular flight.

The consequences of this acquisition are so many and profound that they can only be touched on here, but there is one of them which is so obviously a major evolutionary directive factor that it deserves mention before going further. This is the pollination of flowers by insects, and to a lesser extent by birds and other animals, in absence of which the evolution of many types of flowering plants and, reciprocally, of their pollinators, would probably have been very different, both in course and consequence.

The flight of birds and other animals is so familiar that it often fails to excite the wonder it deserves but it affords one of the prime riddles of evolution, this being how the ability of the animals concerned to launch themselves into the air, and to remain there, hovering or directionally moving, for considerable periods, was achieved. The terse answer is by the development of beating wings; the crux is how these organs have become evolved.

If a *vera causa* for the colonization of the land by plants and animals from the sea is sought, it is likely, as has already been suggested, to be found in some expression of the fundamental principle that 'nature abhors a vacuum', and this in biological terms means the eventual occupation to saturation point of all available habitats. This view involves the assumption that there is, in living nature, an inherent and irreversible urge or tendency towards expansion and fulfilment, using those words in a more general sense. This has been expressed or postulated in one form or another many times and by many people and it is a foundation of what was earlier described as *orthogenesis* but it is also well known in the shape of Nägeli's *Principle of Progression*. This latter supposes that living things are compelled (though obliged is probably a more suitable word) by some innate internal impetus to develop new forms independently, up to a certain point, of the environment and of natural competition. There are various versions or translations of this idea and it may occur to the inquisitive enquirer that something much like some aspects of it has recently emerged under the name of *Parkinson's Law*, and also in the far from perfect analogy of the billiard ball, the course of which is determined partly by its contact with the cushions of the table, which may be regarded as representing the influences of the environment, and partly by the 'indwelling force' imparted to it by the 'stroke of the cue'. This may seem to be a phantasmal kind of explanation but as the subject of animal flight is pursued further it becomes more realistic.

The plants and animals that first colonized the land from the sea must, almost of necessity, have been benthic forms moving about on the bottom of shallow seas or actually attached thereto. For them the transmigration meant, above all, the development of protection against desiccation and may have involved nothing fundamentally new, but the subsequent achievement of flying was quite another matter, especially in the case of the insects. In birds the essential organs of flight, the wings, are modified versions of the two fore-limbs, organs which have existed as long as quadrupeds, and their gradual change is not hard to envisage. Rather similarly the wings of bats are modifications of the fore-limbs, though on a rather different plan, and including to some extent the hind-limbs also. In insects however, the wings are fundamentally different, being structures additional to any found, pristine or modified, in other animals, and produced *above*, and quite independently of, the six limbs. They therefore present the profound problem of what can have caused them to arise. The details of their development are complicated but the more general points can be expressed quite briefly. In those flying insects, such as grasshoppers, which have an incomplete or gradual metamorphosis consisting of a series of relatively small stages or moults comparable with the sloughing of some aquatic crustacea, the wings appear, or at least first become observable, as small emergences which grow more pronounced with each successive moult until they are functional; but in insects with a complete metamorphosis, as is familiar in moths and butterflies, the wings first arise as internal emergences of the inner surface of the skin of the caterpillar and become recognisable as rudimentary wings only in the chrysalis stage.

It would seem likely that differences in development between vertebrates and insects relate to the fact that the former have an endoskeleton and the latter an exoskeleton, but the real question is what induced so many insects to invade the air and what impelled the production, in morphological terms, of an entirely new organ, the arthropod wing. That the first of these questions has closely to do with what has been exemplified as the Principle of Progression may be accepted, but the second is much more difficult to explain. What possible power can have initiated this new organ and designed it to take its place so comfortably among those already existing? Is there or is there not some master plan by which the potentialities or rudiments of the future are foreshadowed from the very beginning and, if there is, who or what determines it?

There is another and rather different aspect of the matter which greatly exercised the early evolutionists and provided one of the most cogent criticisms of the Theory of Natural Selection, and which has never been satisfactorily countered, the question of the value, and indeed the viability, of organs in *process of evolution*. There is every reason to believe that evolution operates so gradually that any new function or any great change in an existing function must pass through early stages during which it is valueless because it is incomplete. This point of view is hard to reconcile with the concepts of 'natural selection' and 'use and disuse', and animal flight is an outstanding example of this difficulty. That is why so much attention has been paid to it here. This subject has, moreover, a powerful attraction of its own because wings, and especially the wings of insects, are one of the most beautiful and intricate manifestations of nature, comparable, perhaps, only with the display of design in the flowers of the higher plants.

The third illustrative example of evolution in the subaerial environment is afforded by that great group of the Angiosperms or Flowering Plants, the grasses. Unfortunately we know little or nothing about the origin of the grasses or of their geological age, except that their widespread occurrence is usually held to date from the Miocene, partly at least because they are unsuitable subjects for preservation as fossils, but there are reasons for thinking that the herbaceous grasses find their more immediate ancestry among the woody bamboos of the tropics. At all events the former today occupy a unique place in the vegetation of the world because their gregariousness produces, at one or other season of the year, a continuous verdant vegetation which may cover great areas and which, in its purest form, is free from any vestige of forest, which is the other great virgin form of terrestrial plant life.

Forest, though in its most luxuriant form teeming with almost every kind of terrestrial life, nevertheless by its very character imposes physical and ecological limitations, especially in freedom and facility of movement, on its inhabitants, and these limitations are generally held to make it unlikely, if no more, that the later evolution of man himself could have taken the course it did in such surroundings. Grassland has no such drawback but provides a more or less completely open habitat where there is little hindrance to the movement of at least the larger animals. Not only this but the actual grass blades and stems provide an easily accessible and wholesome food for many animals. In short, grassland and the closely related form of vegetation called open savannah, affords an environment so exactly suited to the requirements of a great section of the larger mammals, the ungulates or hoofed animals, that it can scarcely be doubted that these have evolved in response to the existence of this kind of vegetation. Certain it is that, until man began to destroy it, the 'grasslands' supported a fauna, principally of such larger animals and their predators, unrivalled in richness and variety.

The steps by which man first became a tool-making and weapon-using animal are matters of considerable debate, but it is usually believed that he first emerged as a hunter, and for this role the 'grasslands' offered not only an admirable field of activity but also quarry in abundance. At least it seems safe to say that, had man in his earlier development been unable to free himself from a forest environment, his later evolution would certainly have been much slower, and probably quite different.

Nor does all this exhaust the credit side of the grasses account. Their fruits, comprehensively called 'grain', are highly nutritious to a wide range of animal life; are dry when ripe; are small in size; and remain viable for long periods, so that they escape many of the hazards which beset many other kinds of fruit and seed. Finally, they are, because of the closeness of growth, and despite the relatively small dimensions of most grasses, produced in great quantities. All these things serve to link them especially with the development of man, and his culture has rightly been called a 'grassland civilization'.

To bring to a close what has been a long and sometimes complicated account of the development of the land biota it may be asked whether it is possible to draw from it any fundamental conclusions about evolution in general. That so many facts and factors interlock so plainly in the story certainly leads to a first impression that there has been, and is, a grand design controlling the evolution of living things but, because the evolution of animals has been consequent upon that of plants, this may well be misleading. In the case of grassland for instance it would seem that the development of this particular kind of land vegetation has been largely one of the results of the well-hypothesised spread of the plants of the land away from the more equable and clement climates of the equatorial tropics, wherever these may from time to time have been, and where it may be noted, the bamboos are most at home, into the zones of more rigorous climates and more varied environmental conditions. It this is so then no concept of plan in this context need be sought, except of course in so far as this general spread is itself according to plan, and this it is no more easy to believe or disbelieve.

Part IV
The Origin of Life

Premises that life originated on the earth and that this planet has, throughout its history, been gradually changing, chiefly in the direction of temperature values . . . the equator-poles gradients . . . oxygen . . . carbon and carbonates . . . carbon-dioxide and carbon-monoxide . . . carbon compounds . . . hydrocarbons . . . methane . . . amino-acids . . . autotrophism . . . photosynthesis and chlorophyll . . . chemosynthesis . . . respiration . . . the ingredient of energy . . . the development of catalytic enzyme systems . . . the 'soup theory' . . . the question of the frequency of the origin of life . . . abiogenesis . . . the possibility that life has originated more than once at comparatively long intervals of time . . . the possibility that life is always originating . . . viruses and their distribution . . . the importance of natural saline solutions . . . brief 'resumé' . . . the problem of the origin of progressive and irreversible evolution . . . newness and novelty . . . the acquisition and role of energy . . . the limitations of molecular diagrams and equations . . . photosynthesis and chemosynthesis . . . the sources of energy . . . entropy . . . control of energy essential for life . . . life-span . . . the rhythm of reproduction . . . the life-spans and cycles of assemblages . . . competition . . . extinction . . . waxing and waning . . . the great problem of whether evolution has been fortuitous or the result of a master plan . . . the likelihood that neither the world as a heavenly body nor the life on it is unique . . . the gradual development of theism . . . the recurrence of baffling problems and their eventual solution by additions to existing knowledge.

The Origin of Life

This book has so far been almost wholly concerned with the changes that plants and animals have undergone during their existence on earth. These are all consequential to the prior issue of how this mundane life came into existence.

One opinion is that life may have reached the earth in the form of some living thing from another celestial body, but this is not a widely held view and, even if it were true, it would only make the ultimate problem more remote, because life must have originated somewhere and at some time, and for purposes of argument it is simpler to assume that life, as we know it, began on our own planet.

Another fundamental premise which, from all the available evidence, can reasonably be accepted, is that the planet Earth has, except perhaps for temporary and superficial interruptions, always been gradually changing with extreme slowness, principally in the direction of temperature loss. This leads to the double corollary that life exists in an inconstant world; and that the general planetary temperature of today is likely to be as low as it has ever been. This idea of life on a world that has long been cooling is fundamental to much biological thought and it is from this point of view that the subject of biological evolution is best approached.

In the light of much that has been written in earlier pages there can be no doubt that water is the primary key to any enquiry into the origin of life and the nature of the evolutionary process in general. Whence the water on the surface of the earth and in its atmosphere came is not wholly clear, but two main sources are usually suggested. One is that it has come from the water of occlusion in silicate rocks which escapes as steam at such high temperatures as accompany volcanic action and later condenses: the other is the condensation of water vapour already in the atmosphere when this has become sufficiently cooled. In both cases it is generally assumed that when this water appeared as liquid on the surface of the earth it was pure, at least in the sense that it contained no appreciable proportion of mineral salts in solution, but this state of affairs can have continued only until the waters began to erode the land surfaces of the world or the beds of the oceans, then unprotected by superficial deposits, and the first products of this erosion must have come from igneous rocks since it was not till much later that the products of erosion itself were re-deposited as the first sedimentary rocks.

A closely related point is that water could not have condensed from the atmosphere or elsewhere until the temperature of the atmosphere and of the surface of the earth had fallen below the boiling point of water, and if, as is assumed, the earth is gradually and slowly cooling, this gives us reason to suppose that the water temperature of the primaeval oceans was higher than that which prevails today, and it seems likely to have been a long time before conditions approaching those of the present were reached.

It may therefore be taken as a basis for further argument that life could not have originated *de novo* in a world completely devoid of water. It is possible, though scarcely probable, that if the only free water on the globe was in the form of vapour in the atmosphere some kind of living things might have made their appearance therein if the temperature was appropriate, but evolution in such a medium could not have gone far and must have given rise in due course to evolution in the waters of the oceans, lakes and rivers, as and when these became formed. At least it can be said that the earliest living things of which there is knowledge from the fossil record were evidently inhabitants, not of a gaseous habitat but of a liquid one, and it may be concluded that the evolution that has produced the present biota must have begun, effectively, in water.

Another point is that if the shape of the world has always, since its solidification, been spherical there must presumably always have been some climatic gradient between the equator and the poles and this being so any crucial point in cooling must have been reached first at the poles. Nothing is precisely known about the distribution of land and sea at such remote times but if there was then some land emergent from the water there must be taken into account the likelihood that cooling would have been more rapid in the oceanic areas and that comparatively soon the equator-poles temperature gradient here would have been more gradual than that on any land that may have existed.

From all this it would appear that before any life could originate some critical temperature figure must have been reached, and although it is not known just what this may have been there are strong hints in the facts that some of the albumins and other high-grade proteins coagulate, with lethal effect on the bodies containing them, at temperatures well below that of boiling water, and that life is destroyed by long exposure to temperatures below that at which water freezes. If, then, the first life to express itself did so in terms of protoplasm it must have done so at temperatures within the range at which water is liquid at ordinary mundane atmospheric pressures.

Though water is of first importance, two other substances, oxygen and carbon, afford valuable clues as to the conditions in which life is likely first to have appeared. The former element is abundant in the mineral world; it occurs in considerable quantities in both sea- and freshwater; it provides a large proportion of the atmosphere; and it is remarkable for its great powers of feeding combustion, which, in chemical terms means that it readily combines with many other chemical substances in exothermic reactions. This oxidation, as it is called, is most important biologically in respiration – the process by which food materials are broken down and energy for other vital processes is liberated.

Oxygen is thus another essential pre-requisite for life and it is therefore very relevant that there is a belief, based on various technical grounds, that there was once no free oxygen in the atmosphere. This is likely to be true enough in the sense that there could have been no mantle of free gases covering the earth until that body had cooled to a certain point, and whence oxygen began to accumulate in the atmosphere is not fully clear, but one view is that it may have been derived in some way from the surface rocks. At all events this is another reason for thinking that life could not have existed before a certain stage in the evolution of the planet Earth was reached because there would have been no free oxygen to help initiate and sustain it.

The element carbon is, in many respects, even more significant. This is today one of the most abundant elements in nature, not only in the form of rock carbonates and in the air, but also, and above all, in living matter, to the extent indeed that what used more commonly to be called 'organic chemistry' is now often termed 'the chemistry of the carbon compounds'. The situation is so familiar that it tends to obscure the difficult question of how this carbon comes to be so abundant in living bodies and in certain mineral deposits. Pure carbon occurs on earth only as diamond or graphite; the carbon in the air is mainly in the form of a very small proportion of the gases carbon-dioxide and carbon-monoxide; and there are some natural carbides, but it would seem that the carbonate rocks are its chief repository. These are the salts of the weak and unstable carbonic acid and occur, mostly as limestones, throughout the range of sedimentary rocks, including the very earliest, but the evidence of fossils shows that even in those remote times comparatively advanced forms of life existed. It is also known that many of the most familiar carbonates, among them chalk, have been deposited either by precipitation resulting from respiration or as the remains of plants and animals which excreted carbonates during their lives. Many carbonate rocks are, however, thought to have been formed by the action of atmospheric carbon-dioxide on silicate rocks of the earth's crust, and this suggests that there may once have been more carbon-dioxide in the atmosphere than there is now.

Carbon-dioxide is readily soluble in water, as is evidenced by the photosynthesis of plants living sub-merged in saline or freshwater, but the origin of this highly reactive gas and its relation to the formation of carbonates is still uncertain though various explanations have been put forward. A particularly notable point is the very small amount of it (only 0·03% of the total volume) in the atmosphere. Since the biota of the land depends on this tiny amount for its maintenance the problem of its origin is important, and the resolution of this may well reveal fresh clues about the earliest stages in the process of evolution.

Carbon combines with one or more of a number of other elements, notably hydrogen, oxygen, nitrogen, sul-phur and phosphorus, to give a host of 'carbon com-pounds' with molecules of a great range of complexity. The simplest of these are the hydrocarbons, composed of carbon and hydrogen only, and these are so important fundamentally that the chemistry of the carbon com-pounds has also been called 'the study of the hydrocarbons and their derivatives'. The simplest hydrocarbon, and one of the most reactive, is methane, which has the chemical formula CH_4, making it particularly susceptible to chemi-cal elaboration.

Methane is most familiar to us as one of the products of the decay of organic bodies, especially plants. It has been detected elsewhere in the universe but it is difficult to find a positive statement about its occurrence as, or in, inorganic matter on earth. If it does so occur as a perfectly free substance quite divorced from living bodies then it might well, because of its great potentiality for chemical com-bination, be the most likely compound from which others of greater complexity could be formed. Some believe that methane was a constituent of the first primitive atmosphere of the earth and if this is so it may have played an important part in the origin of life.

So also may another group of comparatively simple carbon compounds, the amino-acids. There are about two dozen of these and they are closely related to the more familiar fatty acids such as acetic acid, but differ from them in that their molecules contain the amino-group NH_2, and it is this combination of acid and alkaline potential, as it were, which makes them particularly significant. More technical information about them must be sought elsewhere but their importance in these pages is twofold. They are found only in living, or recently dead bodies, as constituents of protoplasm; and they can be elaborated chemically into the simpler kinds of proteins from which, in turn, the higher grade proteins that are the essential ingredients of protoplasm itself can be built up. The amino-acids may thus be thought of as the bricks or building blocks from which the foundations of living matter are constructed.

A consideration of water, oxygen and carbon leads directly to what, in its effects, must certainly be regarded as the most fundamental of all biological subjects, particularly in relation to the problems of the origin of life, autotrophism. This is the name given to the state of affairs in which certain living things are able to manufacture their own food supplies from comparatively simple materials in the mineral side of nature by absorbing them and converting them into more complex substances with which the living body can sustain and reproduce itself: they are able to use chemicals of purely mineral origin in the manufacture of their own food. Each and every fully autotrophic organism is thus completely independent of any other.

One of the most conspicuous features of autotrophism is the unevenness with which it is distributed in the living world. With comparatively few exceptions the members of one whole side of organic nature are autotrophic, namely the characteristic members of the plant kingdom; nearly all other living things are heterotrophic, depending on other living organisms for their sustenance, and by and large these latter constitute the animal kingdom. The only significant complication to this simple picture is that some of the living things, notably the fungi, which, if they were autotrophic would be typical plants, are not autotrophic and are therefore as dependent on others as are animals.

All organisms that manufacture their own food materials in this fashion are able to do so because they produce, in the course of their own metabolism, certain organic substances that play the parts of catalysts in the presence of which some chemically quite simple, but vitally important reactions, unknown in other natural circumstances, can take place.

These particular catalysts are normally associated with the presence of green pigment or pigments known as chlorophylls (though their colour may be masked by other pigments) and they function only in the presence of light, and hence the essential chemical combination which they help to carry out is called photosynthesis. Some very simple organisms, among them certain bacteria, are autotrophic in less sophisticated ways and can obtain what they need for their metabolism more directly from mineral sources by some other process or processes, generally thought to be catalytic, which are given the name chemosynthesis, but green plants are, overwhelmingly, the only living things capable of tapping the huge resources of the mineral world directly and of converting these into food not only for such other plants as lack their own supplies of chlorophyll but for animals also. Hence the natural economy of the world today, and the well-being of its inhabitants, including man, is firmly based on photosynthesis, which is the characteristic expression of autotrophism.

To put this all-important matter more precisely chlorophyll-possessing plants are able, by the agency of their catalysts, to absorb carbon-dioxide from external sources and to combine it with water to form the simple sugars which are the bases of their own nourishment. Part of this sugar is then oxidised in the exothermic reaction of respiration to produce energy, with the aid of which other sugar is built up into new body material. In one way photosynthesis and respiration can be considered the reverse of one another, the former involving the absorption of carbon-dioxide and the liberation of oxygen, the latter the absorption of oxygen and the liberation of carbon-dioxide. The chief practical difference between the two is that photosynthesis goes on only in light while respiration goes on continuously throughout life.

There are two great unsolved problems about auto-trophism which bear directly on the question of the origin of life. The first, and more general, is how such a process originated. Without autotrophism in the form of photo-synthesis there would be no life in the terms familiar today, but photosynthesis is an essential expression of life itself and occurs only in its presence so that it is equally true that there is no photosynthesis without life. This, in broad outline, is one of the most teasing problems in any search for the earliest stages in the origin of life and subsequent evolution and many attempts have been made to resolve it, generally by suggestions that the present state of affairs may not always have prevailed and that long before the evolution of such catalysts as chlorophyll there existed certain entities simpler than, and not strictly comparable, either in form or function, with any living things known today, but this does not explain how such entities may themselves have come into existence. The second and more particular question is presented by the fact that chloro-phyll (and doubtless other catalysts contributing to autotrophism), are in molecular structure far more com-plex than most of the substances on which they operate. The evolution of these sophisticated catalysts must surely have been a slow and ever-elaborating process and in what ways the earlier stages in this succession were able to function, and on what organisms they exercised their powers, is not clear.

From the point of view of evolution photosynthetic autotrophism has another significance not always fully appreciated. Because chlorophyll-containing plants are the only natural makers of food materials, not only for themselves but for other living things too, it follows that some kind of autotrophic plant must have been among the first fruits of evolution and that until these existed no animal life could have developed. This means that animal evolution is consequent upon plant evolution and closely conditioned by it, and the fact that the grand succession of the major groups in each of the two biological kingdoms is parallel in so many respects is striking evidence of this. There is also good reason for believing that, while evolution has been one and the same process in both, the means by which it has been achieved may have been somewhat different. This lack of understanding about the priority of plants, caused in part by an understandable concentration of interest on man and other higher animals, has distorted many evolutionary studies.

It is a reasonable conclusion from all this that the first entities to which the epithet 'living' could be applied must have been able to feed themselves and to reproduce, for otherwise there would be no lineage and the supply of life could have been maintained only by the repeated origin of it. But as far as our experience goes these two crucial activities cannot be performed without the employment of energy, and it must therefore have been possible for these earliest living things to acquire it, and the real question is how they may have done this.

This ability to acquire energy and to conserve a portion of it in the form of food-stuffs is the characteristic feature of the plant world and it is not found anywhere else except, as mentioned above, in those simple organisms capable of chemosynthesis. It is usually supposed that chemosynthetic forms of life played comparatively little part in the development of life as known today and that the great and continuing pageant of developing life on earth began with what would today be called plants, possessed of chlorophyll or some antecedent form of catalyst, but perhaps without all the features now associated with this side of nature. If this reasoning is sound it seems to follow that some of these earliest plants must, at some period, have lost, or failed to make use of, their chlorophyll, and thus ceased to be autotrophic, and it is presumably from these consequently dependent forms that the animal world is most likely to have arisen.

However this may be it must be supposed that, for a long time after the first life appeared, evolution largely took the form of a gradual elaboration of body structure accompanied by the gradual development of increasingly complex catalytic systems, until there emerged those which now go by the name of chlorophyll. At this point a very interesting consideration arises. Why do all existing autotrophic plants, whatever their somatic organization may be, contain this particular group of substances and function by means of them? Why is it that even the simplest of today's green organisms, e.g. unicellular algae, appear at least to possess the same, or at any rate a similarly complex, catalytic system as the most highly developed land plants? The more this point is considered the stranger it appears because many unicellular green algae exist in environmental conditions so different from those of the larger land plants (for instance many of them are plank-tonic) that it is hard to see how one and the same catalytic system can or could suffice for all. If all the possibilities are to be canvassed then one of them must surely be that the simplest chlorophyll-containing organisms of today are not the lineal descendants of the most primaeval plants but have achieved a degree of resemblance to them through some subsequent refinement of evolution.

Last with regard to this discussion of autotrophism in general and of photosynthesis in particular it must be pointed out that, despite a great and varied amount of scientific research, we do not yet know for certain what the catalysts concerned are nor how they operate, and we are not able to carry out the chemical processes involved under artificial conditions and as a manufacturing enterprise. This continuing failure to cope with what must be reckoned among the most important of all scientific problems is all the more difficult to understand because many difficulties which once long baffled investigators have now been overcome by devoting to them very large amounts of money, material and application. This being so it is hard to resist the conclusion that if provision on this sort of scale were made, and a sufficiently powerful, widespread and co-operative organization supported it the problem of photosynthesis could, and would be success-fully resolved in a comparatively short time. At least this may be hoped for because, otherwise, the human race will eventually be faced with starvation.

The facts and assumptions set out so far in this Part form the powerful basis of what is currently the most acceptable view of the way in which life may have originated on the earth. This is generally and con-veniently, though not very elegantly, spoken of as the 'soup theory', and it can be outlined as follows.*

However pure the water that condensed from the atmosphere to form the oceans may have been at first, it cannot have maintained this purity long because many mineral substances are soluble in water, so that as soon as an aqueous erosion of the rocks began it would have been vitiated by the products of this process, and the waters would have become a diluted but no doubt complex aqueous solution. It is supposed that this solution became more concentrated as time went on by the accumulation of salts in increasing variety and reaction until the water contained traces at least of so many substances of so many kinds, either in suspension or in true solution, that it could be likened in fancy to a very dilute broth. As the complexity of the solutes increased so the chemical reactions between them would have multiplied and

---

* What has come to be called the 'soup' or 'chicken soup' theory chiefly derives from the writings of Oparin, A. I., *The Origin of Life*, 1938 and later editions, and from Oparin, A. I. and Bernal, J. D., *The Origin of Life*, 1967.

intensified until at last a stage was reached at which there were formed what, today, would be called 'simple organic molecules'. These, it is suggested, became further consolidated, perhaps under the influences of such external factors as ultra-violet light and electrical charges, into very large and intricate molecules capable of self-perpetuation, and these eventually gave rise to something deserving of the name of 'life'.

If the concentration and complexity of this dilute 'broth' gradually increased, at least for a time, there must be taken into account at least three theoretical possibilities – that the concentration of solutes in sea-water became stronger, though not necessarily continuously so; that at some stage it gradually became weaker, perhaps because more of the substances in solution were precipitated or otherwise taken out of circulation; that the 'salinity' of sea-water long ago reached some kind of equilibrium, to which, it seems likely, the life that had developed in it notably contributed.

At first hearing the 'soup theory' has two virtues which help to commend it. It has simplicity, which needs no further comment; and it is invulnerable, meaning by this that it is so broadly based and generalized that it is hard to deny its validity by producing contrary evidence. Neither of these features necessarily strengthens or weakens it but they do make it desirable to examine its implications with care.

In respect of the structure and sequence of the actual chemical processes involved these implications are many, but from a wider point of view two are of particular importance. The first is the strong inference that, whatever the details of the origin of life in this way may have been, the process as a whole occurred only once and that it was associated with particular conditions prevailing at a particular time, or, to put the point rather differently, that life on earth has come into being on only one never-to-be-repeated occasion. This idea agrees with the traditional accounts of Creation, and how far indeed it is the child of such deeply founded beliefs is an interesting speculation, but at all events that there has been a single never repeated origin of life is now the most widely acceptable of the theoretical possibilities. The second and by far the more profound implication is that the origin of life involved or included a change which resulted in what had previously been inanimate becoming animate, and that the Rubicon between non-life and life was somehow crossed. The notion that the origin of mundane life, *de novo*, within an inorganic environment was a single spontaneous event, unique because it was never repeated, is simple enough, but it conceals a number of uncertainties and needs careful examination. In its very simplest and strictest expression it can be taken to mean that at some precise moment a single complex carbon-compound molecule became activated and so became the first *organism*. In a wider interpretation it can be held to mean that, within a brief period, life appeared in various individual molecules, but that this process eventually ceased. This rendering can be put differently by saying that before a certain point in time the world was lifeless: that for a while thereafter life in one or other form made its appearance, but that this process ceased comparatively soon and has never been resumed.

The difference between these two interpretations may seem of little consequence and leaves the general hypothesis of the 'soup theory' unaffected, but in fact it is of fundamental importance for the study of later evolution because it is only in the former circumstance that it can be accepted as beyond dispute that all the members of the two biological kingdoms are, and always have been, the genealogical blood relations of a single lineage. Even if there were only two closely coincident but separate origins of life it would be impossible to make this assumption and the study of biological relationships would be far more difficult even than it is now. However this does not affect the principal tenet of the 'soup theory' which is that life originated on earth over a given shorter or longer period of time; that this period followed the beginning of the erosion of the earth's surface rocks; and that this eventually terminated with the cessation of any further life production *de novo*. The second crucial phrase in this statement is 'once and for all' and it is legitimate to wonder how it became so firmly an ingredient of the theory because there is no *prima facie* reason for it. It seems likely that the idea stems mainly from three sources. One is that in a world whose history is one of ceaseless change conditions suitable for the initiation of many kinds of biological processes are likely to have occurred and to have changed as time passed; but this is not to say that every phenomenon must sooner or later come to an end, and indeed the very long-continued existence of life supports this opinion, though it is impossible to say how near its absolute termination life on earth may be. Another source, already touched upon, is the lingering tradition of the deeply ingrained story of Creation as related in the *Book of Genesis*. The most realistic and logical source, however, is almost certainly a disbelief in spontaneous generation, or 'abiogenesis' as it is more scientifically called.

Today the situation regarding spontaneous generation is of great interest. For many hundreds of years it was generally believed that living matter could appear spontaneously in suitable conditions, and such processes as the putrefaction of meat and the appearance of maggots in it, and the fermentations of certain organic substances, were accepted as examples of this. After the bacteria were discovered, with the aid of the microscope, attention became focused on the simplest living things and it was commonly believed that they too were the results of spontaneous generation. This is the most important point in the story, because, when Pasteur, in the middle of the nineteenth century, established that putrefaction and fermentation did not occur in the absence of bacteria, that is to say in the absence of any *living* things, the conclusion was rather jumped at that *all* life came from pre-existing life and the idea of spontaneous generation fell into a disrepute from which it is only now beginning to reemerge. This is because of a curious misapprehension about Pasteur's work. He did, indeed, demonstrate that life did not originate in certain conditions which had hitherto seemed to indicate that it did, but he did *not*, nor ever could, show that life could not originate under *any* conditions, Apart from anything else all the evidence goes to show that there was a time before which there was no life on earth and therefore that it must have originated at least once, and the general idea of 'once and for all' in the 'soup theory' looks very like a combination of this knowledge with what was originally concluded as a result of Pasteur's work.

Although the 'soup theory', by suggesting that life on earth began at a particular time when special conditions prevailed, invites the conclusion that it was a 'once and for all' event it does not exclude other explanations, and to these attention must now be turned. The remarkable unity of life throughout both biological kingdoms has already been stressed, and rightly so, but when the whole wide range of structure, form and function therein is realised the more difficult it becomes to believe that all these living things are related directly to and descended from a single original form of life. This hesitancy is also as it should be because there are two other obvious possibilities that must be taken into account.

The first of these is that life has originated on the earth more than once at comparatively long intervals of time. How many times it is impossible to say, except that if the intervals were very long they must have been few in number in order to be accommodated within the span of geological time as a whole. The fossil record certainly reveals 'inconsequences' and 'unconformities' which it is hard to account for except on the hypothesis that they represent periods of time which have left no evidence of their passing, but what lengths of time these gaps may represent it is impossible to say. It is equally impossible to be sure of what may have happened during them or what the conditions of their times may have been, but apart from this there is no logical reason why a rarely repeated origin of life should interfere with the continued evolution of the plant and animal types already in existence, in which case the re-emergence of life at long intervals would effectively be periodic reinforcements of the living stream. A more significant aspect of the matter, however, is that it would in some measure explain what has often and long puzzled students of evolution, namely why there are so many grades of existing plants and animals and why these include some at least of extreme simplicity, representing, if our interpretation is correct, some of the earliest forms of life, because according to any hypothesis of repeated origin these different types could represent the respective contemporary culminations of lines of evolution whose beginnings were at different times. From this there arises the interesting reflection as to whether, if the chasm between non-living and living has been crossed more than once subsequent evolution has always followed the same course. To suppose that it has not would lead into a quagmire of speculation, but this is no reason for dismissing the idea out of hand. However it would lead far from more immediate problems of evolution, and to keep the argument within reasonable bounds it must be supposed that any repeated origins of life would all have been the same, or at least that the same kind of synthesis of organic from inorganic happened more than once and has on each occasion been followed by like results.

On this assumption there is no reason why the present state of the organic world should be thought less compatible with a repeated origin of life than with one only. Indeed many may think that it is more so because the major groups of plants and animals are, on the whole, sharply defined. True, there are some forms which lie between them but these are less in evidence and much less justifiably recognisable as connecting them than might be expected. This general absence of intermediate forms (or 'missing links' as they are often called), which is based on the discontinuity between assemblages described in Part I, has long been a matter of interest to students of evolution, and it certainly suggests that some of the larger groups of living things may either have branched off from the main stem of the ancestral tree much earlier than the others with which they are most often associated; or have followed their own independent development from the beginning. In either case what would appear to be intermediate or connecting links should be regarded rather as the products of more or less independent evolutionary lines of various ages characterised by features which, in terms of man-made classification, place them somewhere between groups with which they have little or no real affinity.

But if the possibility is allowed that life may have originated more than once then a new consideration emerges. A quantitative element is introduced and the argument must take into account the frequency of the event. The common belief in 'once', which arises largely from the supposition that the conditions for it must have been very particular and unusual, tends to encourage the belief that repetition, if it has occurred, must have been at long intervals of time, but there is no obvious reason why this should be so, and if precisely similar conditions were requisite on each occasion the likelihood that these were widely separated in time is much diminished. More important is to consider what 'more than once' may mean. Does it imply twice, thrice, several or many? The first three of these possibilities have been covered, but much more remains to be written about the last.

Many is a very indefinite word, but in relation to this discussion of evolution its logical extreme is that life has originated so often that the process has been virtually or actually continuous, always taking place somewhere or other. This idea that life is continuously coming into being within the mundane environment is not, at first, likely to commend itself readily, but it must not be dismissed lightly. It is no more inherently unlikely than other ideas, because there is no yardstick by which their relative probability can be assessed; nor is it more or less likely than any other suggestion because there is no direct evidence for any of them. We are ignorant of the essential facts and all we can do is to make suggestions based on what *is* known, or assumed to be true, of the natural world in general and of organic nature in particular. In short there is no less degree of probability in the notion of a continuous origin of life than there is in the idea of a single origin or in origins repeated at long intervals of time, and, indeed, if the matter is considered dispassionately a continuous origin is the more promising of the three, if only on the grounds that a process of origination once activated would, *as long as its products went on to develop and thrive, thereby demonstrating that the conditions necessary for life continued to prevail*, itself be more likely to continue than to stop, never to be resumed.

If the conditions necessary for the origin of life are as particular as the 'soup theory' infers then change in them might have one or other of two rather different effects. On the one hand it might inhibit the continuing origin or renewal of life while nevertheless making the environment *more* rather than *less* favourable for the perpetuation of life already in existence. On the other hand it might make the continued initiation of life all the more probable, especially if, as has been suggested, it included the enrichment of the environmental supplies of carbon-dioxide, oxygen and other important substances as a result of increasing biological activity in the world.

It is interesting also to consider whether such a general enrichment may not have influenced the precise course that evolution has followed and, particularly, whether it may not have been the cause of the increasing differentiation and complexity that is one of its main features. Certainly it is difficult to suppose that such an enrichment, which would be exceedingly slow, could be a hazard to life in general; rather it would seem much more likely to be beneficial to it, and the very multifariousness of living nature today, especially on land, points in this direction. It is considerations like these that make the idea of a continuous origin of life more credible than it might at first seem.

A main objection to the idea is the lack of any evidence in the contemporary scene that it is happening. We cannot point to this or that fact and say 'here is life in the course of independent emergence', but is it likely that we ever could? This is a very difficult question to answer because it is uncertain whether or not we are as yet fully acquainted with the very simplest forms that life can express. The bacteria, though their existence has long been known, have been deeply studied only during the past century or so, and the viruses for only about half that time, and there may still be much more to be discovered about them.

The viruses, some of which are all too familiar as the causes of human ills, are so small that they are commonly measured in millionths of a millimetre and the smallest of them compare in size with certain complex organic molecules, but even so some of them are described as having a quite differentiated structure. They consist of units which, in suitable surroundings, are apparently capable of some sort of replication and self-perpetuation, but this is only at the expense of the cells of certain plants and animals and, it would seem, by methods not altogether the same as those of reproduction in most living things. This is why, in brief accounts of the viruses, authors are reluctant to apply the words 'life' or 'living' to them and content themselves with describing them as infectious 'agents', 'agencies' or 'particles'. Their close association with plant and animal cells is particularly noteworthy because it is difficult to understand how and where they may have lived before the existence of their present hosts. One particular kind of virus, 'phage' or bacteriophage, attacks and destroys living bacterial cells only, and this may be especially significant because it seems to show that the phage can exist outside the bacterial cells, at least for a time, as indeed must be the case if it is to spread among huge numbers of unicellular organisms, and this may be of considerable importance in relation to its nature and evolution. The simplest plant-infecting viruses compare with the phages in structure and, together with more differentiated kinds, occur only in the cells of certain plants and animals. Here they are often destructive, breaking down the protoplasm with deleterious effect, but most viruses do this only to a limited extent and live as obligate parasites dependent upon the survival of their hosts, and they are able to pass from host to host in some free manner.

The viruses also appear to have a remarkable distribution among the classes and phyla of the plant and animal kingdoms, though this impression may be partly due to lack of knowledge concerning certain groups. The most notable feature of this distribution is its unevenness, and here three points are particularly noteworthy. First, there are various large groups of living things in which viruses have scarcely or not at all been detected. Second, these groups include most of the marine invertebrates and, in particular, the brown and red seaweeds, though some are known in the green and blue-green algae. Third, the very great majority of viruses are found in three great sections of the living world, namely the seed-plants, the Arthropoda and the Vertebrata, and these are the three chief contributors to plant and animal life on land. There is also a small occurrence of viruses in one or two other groups of which a minority of species inhabit land or freshwater. It certainly seems that there is some connection between viruses and organisms of the land; and an absence of such association between viruses and the inhabitants of the sea. To discuss these points further would be too much of a digression but it is certainly worth while to remind the reader that one of the general distinctions between marine plants and animals and those of the land is that the latter have some kind of exoskeleton or dermis which protects them in some degree from the loss of body fluid to the atmosphere by evaporation. Whether this, which has the effect of imposing a barrier between the protoplasts of the organisms concerned and the external environment, has anything to do with the distribution of viruses among the main groups of living things is at least an interesting speculation.

Consideration of the viruses leads on not unnaturally to another difficult point in the problem of the origin of life, namely that the least complex living bodies are, in chemical constitution, vastly more elaborate than any molecules of the inorganic or mineral world, and yet, if the general picture is reliable, the difference between the two must have been bridged at least once by intermediates, that is to say by molecules more functional than any in the inorganic world but less so than those which compose living matter. How, when and where could these have occurred? However many intermediates there may have been the more advanced of them at least must have contained some of the elements present in the proteins that constitute so much of what is called protoplasm. But all the more complex organic substances known today are found either in living bodies or in the products of the death and decay of these and the idea that what may be called 'highly elaborate carbon molecules' can exist wholly apart from organic bodies either alive or dead is not in accordance with contemporary scientific thought, though there are claims that such have been detected. Yet presumably they must have so existed at some stage if life was ever to come into being?

In one way it is of course true to say that all biological reproduction is one aspect of the origin of life, and the common use of such expressions as 'a new life' reflects this. It can even be argued that since the nutrition and assimilation on which reproduction is based is, at bottom, the synthesis of carbohydrates from carbon-dioxide and water, both of which are normally regarded as constituents of the mineral world, reproduction is in some respects a constant repetition of the origin of life, but this is too facile a line of reasoning because although new proteins and other ingredients of protoplasm are formed in the process, their vitality is but a projection of the already-existing life of parents. Man has long been conscious of this continuity of the life-force down the generations, and at one time the words 'continuity of the germ plasm' were commonly used to describe it.

Any idea that life is, in the strictest sense of the word, continuously originating is also unpalatable to most people today not only because disbelief in abiogenesis is still strong but also because there is no indication where, if life is indeed originating, the process goes on. Certainly not everywhere because the liquidity requirement alone would prevent this, and if the common opinion about the marine origin of life is substantially correct then the field can be narrowed still further to those natural situations where there is water containing a mixture of salts in solution similar to that found in the sea. Clearly this means in effect the sea itself and it is here that life is most likely to be found arising *de novo*, but it is here also that any evidence of this happening is likely to be least discernible. The land, in nearly all its expressions, can be continuously and minutely kept under observation but the sea is quite a different story. It cannot be penetrated and observed in the same way and the very vastness of the oceans, and the inaccessibility of their depths, mean that great parts and many aspects of them are still comparatively unknown. At least we may be sure that the deep blue sea has still many secrets to reveal.

As regards the land surfaces of the earth or, more precisely, of the subaerial environment, the problem is different because the distribution within it of quantities of saline aqueous solutions such as can be compared with sea-water is very limited. In a few large inland waters the concentration of salts is as high as or even higher than it is in the oceans, and these (of which the Dead Sea is the most familiar) are generally called seas, but their waters cannot be regarded as strictly comparable with those of the oceans proper not only because the concentration of salts in them is likely to vary considerably by evaporation but more important because of the absence from them of wide-ranging tides and their many biological consequences.

From this it would at first seem that the kind and quantity of saline solutions in which, according to theory, life is most likely to have arisen do not occur, at least in any but sporadic and occasional form, in the subaerial environment, and that in consequence the idea of any origin of life there can be precluded, but there is one remarkable reservation to this. One of the reasons why it is believed that life began in the oceans is that the body fluids of plants and animals, e.g. the liquid in the cell vacuoles of the former and the plasma, lymph and so on of the latter, have a chemical constitution at least reminiscent of that of sea-water, and this has been taken to indicate that these liquids are, as it were, a continuum of the ancestral medium in which the first organisms were immersed and began to evolve. Whether this is unduly fanciful or not the fact remains that the body juices, including the contents of most cells, of plants and animals living on land or in freshwater provide what is, in total, a considerable reservoir of saline solutions – solutions which, be it noted, are by the very nature of the case intimately associated with active life. Is it mere coincidence that the viruses are so often found in these body fluids?

It is not the present purpose to argue the relative merits of the three possibilities of the origin of life further but the discussion of them may be concluded with a final brief word about each, based on the generally accepted belief that the earth as a potential or actual habitat has been continuously undergoing changes since its earliest beginnings. If the origin of life, or more precisely the transition from inorganic to organic, can take place only under particular conditions then the first possibility, once and for all, is not only the most likely but also the least open to contradiction. If the same, or a closely similar, transition from inorganic to organic is a necessary precursor of life then, because of the general change with time in the environment, it seems improbable, though not wholly impossible, that life has originated more than once at different stages in the earth's history. One side of this argument applies also to the third possibility, that life has never ceased to come into being since its first inception, but here there is the difficulty of distinguishing between the original act and its continuation. If, as seems likely, the environmental conditions have, over the ages, become more rather than less favourable for the maintenance of life, then the probability of a continuous origin is greater. In short the choice seems to lie between a 'once and for all' origin and a 'continuous' origin. At present the weight of informed opinion, could it be properly assessed, would almost cerainly be in favour of the former, not least because that is the traditional view, but circumstances and beliefs are just as subject to change with time as everything else and it would be unwise to discard the possibility that further knowledge may lead to a different conclusion.

Deep as the puzzle of the origin of life may be in terms of chemistry and physics there is one feature of the living things that eventuated from it which is even more difficult to understand and which is one of the ultimate problems of biology. This is the question of how and why living organisms became endowed with that faculty of progressive and irreversible change of form and function that is called evolution? Change in itself is not the prerogative of the organic world as any study of geology shows, but this latter kind of change partakes rather of the kaleidoscope in which the effects, however multitudinous they may be, are all produced by the rearrangement of a number of pieces which are not subtracted from or added to from outside. In organic nature however the changes are not changes of mere rearrangement but changes which effectively convert the matter concerned into something different in its very nature. It is difficult to choose exactly the right words here but perhaps to say that living things have the facility of changing matter into forms hitherto unexpressed gives some impression of the peculiar part they play in the world. It is this 'novelty' or 'newness', this constant production of something never before seen or experienced, as opposed to the rearrangement of familiar patterns, that is the crowning distinction of life and which sets it apart from inorganic nature.

In a biological context there are two aspects of 'newness'. One is the appearance, in the course of evolution, of characters hitherto unrepresented; the other is the initiation of life itself. The former, round which the great science of genetics has grown up is much easier to discuss because it has, for that reason, a tangibility that the other lacks. Much is known, and can be presented in diagrammatic form, of the machinery involved in the inheritance of characters and of the ways in which these may change from generation to generation, and there is a great body of literature illustrating, and to some degree explaining, this. A good deal is known about how these things come to pass but much less about *why* they happen. It is nearly always easier to suggest how an event occurs than to explain why it happens, and the subject with which the science of genetics is concerned is no exception to this, so that however much admiration there may be for the achievements in this field it must not be allowed to blind us to the fact that whatever may be known about the fate of characters after they appear, or about the mechanism by which they change, there is still much to be learnt about why new characters take the mould they do and so steer the ship of evolution in particular directions.

The greater expression of 'newness', the initiation of life, some aspects of which have just been discussed, is particularly difficult to deal with because it is, almost by the very nature of things, hard to picture what the mundane environment can have been like in pre-organic times, and especially to realise its *lifelessness*, but the more we try to do this the plainer it becomes that the appearance of life on earth was an event of incalculable significance. It altered the world fundamentally in kind, and also in potentiality. If we could dissociate our minds entirely from the idea of life and imagine ourselves as conscious only of the inorganic, such a possibility as life and its consequences would be beyond our comprehension, because we should have no pre-cognition or preconception of such a development. This being so the question of what may have triggered off so astounding an event is a crucial one leading straight to the ultimate biological enigma of how what is commonly called the 'vital spark' was ignited.

This fundamental question takes us back to the 'soup theory', which pictures the origin of life as the culmination of a slow and gradual process of increase in molecular complexity until there appeared molecules capable of *continuous self-perpetuation without diminution* – a phrase which is one of the most satisfactory definitions of the basic vital activity. Whether this picture is correct or not it is certainly not so simple as it may at first seem to be, and this for at least three reasons.

The familiar molecular diagrams of the chemistry books give remarkable representations of the ways in which such simple carbon compounds as carbon-dioxide and methane may gradually be elaborated into the nucleo-proteins found in protoplasm, but they do not explain how, or at what point, this elaboration becomes associated with life. As far as is known no single carbon compound, however complex, in isolation possesses vitality, which only manifests itself when different kinds of such compounds become associated and organised into the physico-chemical system that is called protoplasm. It may be concluded therefore that life comes about, not simply by increasing chemical elaboration, but by the organization into a peculiar system of some of the products of the process. It is this organization, using that word in its meaning of co-ordinating and preparing for activity, that marks the difference between what may be called, in colloquial phrase, 'dead and alive', and of the exact details of this organisation we are as yet not fully aware.

A second point is that we do not know how abruptly the evolutionary transition from non-life to life may happen. If the process is as gradual as is commonly supposed there must be envisaged the possibility that there have existed over a considerable length of time entities which, structurally at least, lay somewhere between the inorganic or mineral and what would now be termed the organic or biological. Such a state of affairs is not hard to imagine in terms of structure, but function is another matter. In what way could these intermediate states have been functional and, if they were, what functions could they have performed? To put the essential point rather differently, is it possible to regard certain stages in organic development as being more completely alive than others? If this is possible at what point can life truly be said to begin? These questions are not so hypothetical as they may seem, and it is interesting that it has actually been suggested that some of the simplest viruses find a place in the shadowy realm between non-life and life.

A third point is nearer the heart of the matter. If a complex molecule is to become capable of self-perpetuation without diminution it must somehow acquire enough energy to enable it to build its replica. It must acquire the ability to use energy for its own purposes, and this is something that what may be called 'ordinary' molecules cannot do. Looked at in this way life is essentially a utilisation of energy for the ultimate purpose of replication.

Here it is difficult to resist using the imperfect but nevertheless vivid analogy of cake-making. The first step in this is to assemble the ingredients, something which may be compared with the thickening stage of the 'soup theory'. Then comes the mixing of these ingredients, the purpose being to arrange them in a three-dimensional system. This done only one thing remains – to bake the mixture. It is here that the analogy is both strongest and weakest; strongest because the mix will not turn into a cake unless it receives a supply of energy in the form of heat from outside, weakest because the injection of energy is in no way conserved for the baking of more cakes but is rapidly dissipated.

This cake-making comparison is useful in another and perhaps more significant respect. The energy which bakes the mix and turns it into a cake comes from a great external store – from combustible fuel; from natural or manufactured gas; from electrical power – and all these have behind them a vast background organisation, and it is certainly permissible to wonder whether there may not be some comparable kind of injection from without by which the gap between the mineral and living worlds have been and still is being crossed. In other words how certain is it that the bridging of this gap is achieved wholly within the limits, resources and scope of a natural environment controlled by the laws of physics and chemistry as they are known today?

This point can be carried further by returning to the subjects of photosynthesis and chemosynthesis. The conundrum about photosynthesis in relation to the origin of life is that although it is essential for the continuation of life it takes place only in bodies which are already alive and in what can only be described as a comparatively advanced state of chemical complexity and organisation. Thus there seems to be a kind of 'hen and egg' situation in which it is not clear whether photosynthesis in absence of life or life in absence of photosynthesis came first, and it is partly at least as a means of resolving this that the various hypotheses relating to chemosynthesis have been advanced. These suppose that the first vital processes deserving that name were much less sophisticated than those associated with photosynthesis and were expressed in bodies much less highly developed than any of those in which photosynthesis is known today. This may well be true, and the very complications of photosynthesis argue that it is far removed from any primaeval condition, but theories of chemosynthesis do not solve the problem, they merely push it further into an increasingly obscure background, because however simple the entities and the processes they conduct may be, the *latter*, if the *former* are to replicate themselves, must at some stage have acquired a sufficient amount of energy, presumably from outside themselves, because without it they could not be effective. Thus there remains the question of how the energy requirements in the origin of life were met, and there seems no escaping the conclusion that evolution, in biological terms, began with an infusion or accumulation of something to which the name energy is commonly applied.

The value of molecular diagrams and equations has been pointed out, but this value is limited and the pictures they show are incomplete because they are over-simplifications in two important respects. Certain vital processes, such as respiration and the synthesis of sugars from carbon-dioxide and water, can be expressed by very simple literal chemical equations and may thus be thought to involve correspondingly simple reactions but in fact they take place effectively only because organic catalysts, many of them enzymes, are present and active. Respiration for instance needs the enzymes called oxidases: the synthesis of sugars normally needs the presence of the organic catalysts broadly called chlorophyll. These catalysts may or may not all fall within the definition of enzymes but the important thing about them from the point of view of the origin of life is that they are credited with a chemical complexity much greater than that of the substances on which they operate. Sugar synthesis, for instance, is, in terms of molecular exchange, a simple enough process, but the chlorophyll on which it depends includes molecules comprising several elements and some dozens of atoms.

This difficulty is resolved, in theory, by supposing that the first living things of mundane origin, which, it is generally agreed must, in absence of free oxygen, have been anaerobes, were able to function by means of much simpler catalysts, but this does not really help. Respiration, aerobic or anaerobic, is pictured as involving very simple chemical interchange and it is difficult to imagine that the catalysts required for it could have been equally simple, and there is thus the situation that the most primitive of living things must have been equipped with chemical substances more, and perhaps much more,complex than those which figure in some simple equations. So it would seem that in order to carry out these apparently simple vital processes there must already have been developed catalysts of more complicated structure, and from this it would seem to follow that before some of the simplest manifestations of life could express themselves there must have come into existence comparatively intricate catalysts, but how these can have appeared in a pre-biological world is hard to understand.

The other important respect in which diagrams of chemical reactions are imperfect is that they do not convey plainly enough that the essential part of most chemical reactions is not so much the exchange of molecules and atoms as the absorption or liberation of energy. Without this energy interchange there cannot, to the best of our knowledge, be life and how such a process came into existence is the great problem.

There are various sources of energy available in, and to, the mineral world – solar, atomic, atmospheric, gravitational, volcanic, hydraulic and so on – but the action of them all is subject to the operation of the Second Law of Thermodynamics, according to which such energy eventually becomes irreversibly degraded and unavailable for further constructive use. The measurement of the degree of this degradation, or unavailability, of energy is called entropy, but this is a very difficult concept to describe simply and it must be enough to say here that the factor to be reckoned with is that the supplies of energy in the mineral world are gradually becoming thus degraded and therefore unavailable to do work. Fortunately such energy sources as the sun and atomic disintegration can, for all practical purposes, be regarded as virtually in-exhaustible resources by drawing on which the full effect of entropy is postponed. Nevertheless, were it not for one circumstance total entropy would eventually result; the availablity of energy would move towards zero; and the 'clockwork' of the world would run down unchecked. This circumstance is that part of nature in effect defies the operation of the Second Law of Thermodynamics in so far, not only that it is capable of capturing energy but also of delaying the effect of entropy upon it so that its degradation is, as it were, constantly deferred. This section of nature is that part of it that comprises living autotrophic plants.

From this point of view the picture of the 'vital spark' looks rather different. The every-day meaning of these two words strongly suggests some instantaneous injection from without by which the whole properties of the matter concerned become altered, but what has just been said about energy indicates that the essential part of the endowment of life may have been something less vivid, and the possibility that the origin of life consisted, *in toto*, of two stages or phrases, one comprising the control of energy and its dissipation, the other actual vivification. If this is so then the latter may be something to which the words 'vital spark' or 'breath of life' are not as appropriate as is commonly supposed.

Analogies are seldom complete but they have their values nevertheless and that of cake-making can, as has been shown, be used to illustrate the point that the *application* of energy is essential. This can be extended to represent the *control of the dissipation* of energy by likening it to the gathering together and mixing of the ingredients in a confined space, the kitchen bowl, from which they cannot escape in all directions. Once under this degree of control they can be modified or, by the addition of fresh materials, elaborated in all sorts of ways. It is even tempting to think that the control of energy is the whole process in the origin of life and that once this has been accomplished the rest necessarily follows, but here the very limitations of the cake analogy help, because it is clear from all experience that whatever may subsequently happen to the cake-mix it does not result in the production of more cakes each equivalent to the original, as is the case with the replication of living things. It must be assumed therefore that there was, at some exceedingly remote period of time, something over and above the formal laws of physical energy (at least as they are known today) which can be described as the conferment of life. Whether it is possible, at present, to probe further than this is doubtful but it is possible at least to consider what some of the immediate consequences of this conferment may have been.

At this point it is convenient to sum up what has been written here about energy in the simplest terms. Radiant energy, so-called because it reaches the earth in the form of light beams, comes in great quantity from the sun, falling on whatever surface is exposed to it. Where there is no vegetation, planktonic or other, this radiant energy goes to warm the materials which it strikes, but for a number of reasons, among them the alternation of light and dark, which result from the rotation of the earth, this heat easily becomes dissipated and is not accumulated. Only when it falls on plants containing chlorophyll does this not happen because it is only then that some of this energy reaching the earth is absorbed and converted into other forms of energy which enable their vital processes to continue until, from one cause or another, the plants concerned die.

Further, if, as our knowledge and experience both go to show, slow change with time is the fundamental characteristic of nature, then anything that serves to direct this change and to make it irreversible, must result in what can only be described as evolution, and the peculiar energy economy of plants seems to provide this steerage.

Although what has been called here the arrest, or postponement, of entropy results from the energy-absorbing activities of plants its effects are not confined to these but are communicated to animals through their dependence on plants for food, and it is therefore a feature, either directly or indirectly, of both the plant *and* animal kingdoms. Not only this but it operates in a very remarkable way. In the sense that it does not cease as long as green plants exist the process is continuous, but this continuity is not an even progression. On the contrary each and all living things have a definite, and usually fairly sharply defined, period of existence which is usually referred to as their 'life-span', a phrase which it is well to use in order to avoid confusion with the more important connotation of the word 'life'. The duration of life-span varies greatly according to the kinds of organism concerned, and there are several particular complications such as the case of insects which pass through the stages of larva, pupa and imago, and that of the whole range of plants and animals which exhibit dormancy, varying from the hibernation of many animals to the exceedingly long dormancy of certain seeds, but by and large the duration of life-span is an assemblage characteristic. It is difficult to be sure in what organisms the span is least, and even here it must be long enough for maturity and reproduction, but from this end of the scale there is almost every value up to what must be regarded as the longest-living of all things, some of the largest trees, in which the life-span is measured in thousands of years. Even these eventually perish if only by accident or disease but, barring this, not until their characteristics have been transmitted to their progeny. Death may be long deferred but it is inexorable. There is nothing quantitatively comparable with these plants among animals, at any rate among those that are mobile and show normal individuality, and in this kingdom one hundred years would seem to be about the longest life-span.

A point of great interest here is the relation between age and reproductive ability. It is true to say that plants, in general, continue to reproduce sexually, at appropriate intervals depending on their growth-form and the latitudes of their natural geographical ranges, as long as they live, and in addition to this some form of asexual multiplication is not uncommon among them, as, indeed, it is in some animals.

Annual or, more strictly, monocarpic plants fruit only once and then die, the race continuing in the form of seed. Such plants are especially associated with strongly seasonal climates, and how common they are in feral equatorial vegetation is not easy to assess. Biennials, which require two growing seasons from seed to fruit, are chiefly associated with strong rhythms of short growing seasons and long winters. Perennials, when growing within their natural ranges of tolerance, continue to reproduce at intervals as long as they live. Tropical rain forest trees, in particular, which live under ample conditions of temperature and moisture, would seem to have an almost indefinite life-span, terminated only, in absence of disease, by accident, but the weaknesses conducive to this may be due to old age.

The corresponding situation in animals is less easy to describe because of their closer and more varied inter-relationships, especially those expressed in social organization and in the predator-prey correlation, but there seems little indication of separate terminations of reproductive ability and of life itself in the individual, though this impression is largely based on the observation of domestic animals or others kept in the unnatural conditions of captivity. That the evolution of the distribution of sex, and its consequences, is still proceeding is indicated by the fact that menstruation is found only in man and the higher apes, other mammals having oestrous cycles in which the periods of 'heat' are separated by comparatively long periods of time, sometimes as much as a year or more.

This matter of life-span is a crucial one because, were it otherwise the course of the progressive variation which is the *sine qua non* of evolution would almost certainly be very different. As was stressed earlier this variation expresses itself most significantly in the differences between parents and offspring, and this being the essence of the system, it is clearly greatly to its benefit that the phenomenon of life-span is such that, broadly speaking, individuals of only a few generations of any particular assemblages (in the more familiar examples perhaps only three or four) are alive at any one time. This means that the newer generations sooner or later have the field to themselves, and can express themselves in ways which would be prejudiced if members of a large number of prior generations continued to cumber the ground. In short, a definite life-span means that at suitable intervals there is what amounts to a 'clearance sale' which leaves the field open for the younger and, presumably more vital members of the population. At the same time life-span is usually long enough to ensure that, in the event of accident to one generation, the assemblage concerned has another opportunity of making good the damage.

Just as every individual has a potential life-span which, broadly, is characteristic of the assemblage to which it belongs, so also every different assemblage has a life-span of its own. This is longer than any individual life-span but it follows the same general pattern. First, a new assemblage makes its appearance as a result of one or other, or of a combination of, the various processes outlined at the beginning of this book. At first it makes little impression on its contemporary society but after this period, which may justifiably be regarded as its youth, it achieves a more active state during which it not only much multiplies its individuals but also, largely as a consequence of this, increases its geographical range. It is during this period of greatest activity, or maturity, that it spawns new assemblages, one or some of which will, in the ordinary course of events, ultimately supplant it. This period of maturity may last for a very long time but in the end the assemblage inevitably begins to diminish. This diminution evinces itself chiefly in geographical contraction or fragmentation caused by failure to maintain itself in sufficient numbers throughout its range, and this decline gradually becomes more and more effective until the last few individuals perish and the assemblage thus comes to have no geographical range at all, and it becomes extinct. It is not to be supposed that every assemblage passes through all these stages at the same rate but it may be accepted that this is the general pattern of life through which all assemblages go.

In the life-span of both individuals and assemblages there is one feature of vital evolutionary importance, the fact that both, throughout their period of greatest activity, produce, at longer or shorter intervals, individuals which differ from their parental units and which themselves tend, under the influence of irreversible variation, generally to diverge more and more from their progenitors. In the case of individuals the life-span is so short that, except perhaps for very sudden and drastic mutations, it is over before the divergence can manifest itself in the form of competition, but with assemblages the life-span is usually so long that many divergent offspring not only come into existence during it but continue to evolve in company with their parents. These later products of evolution, or more modern versions as they may be thought of, enter the general ecosystem and cannot fail to be in some degree in competition with their parents, and this must certainly be reckoned as possibly contributing to the decline of the latter. At the same time this competition must not be over-rated because the new assemblages are likely, from the very fact of their divergence from their parents, to show toleration ranges of their own which may or may not conflict with them. Nevertheless, since the resources of both space and materials are not infinite it follows that the older assemblages are, in course of time, supplanted by some at least of their own younger and, it may be supposed more vigorous, offspring.

The extinction of assemblages is something that calls for particularly careful consideration because, owing to the immense time dimensions involved, there is no practical experience of such an event or of the exact form and course that it may take. The time factor is so long that the incidence of extinction can clearly occur only at compara-tively long intervals, and judging from the usual estimates of the time that man himself has existed it may well be that there has not, in this time, been more than a few instances of extinction as a consequence of entirely natural causes wholly unaffected by the evolution of man. Consequently what we know in practice about extinction is derived entirely from the artificial effects produced by man himself. Here the ingredient of super-abundant time is absent and is replaced by the old Adam of human desire and avarice, and to draw any but the most superficial comparisons between the two is, to say the least of it, dangerous.

Like individuals and assemblages, still larger groups have their own life-spans, but generally speaking the larger the group the less exactly is its life-span prescribed. The fossil record shows this very clearly. All the major groups of living things appear to have begun almost tentatively and to have passed at first and relatively slowly through a stage which may be compared with the infancy of an individual; then to have experienced almost suddenly a great expansion which can be looked upon as representing maturity and maximum vitality; and finally gradually to have diminished, by the increasing extinction of their members, to near vanishing point in a process that can properly be described as ageing. In individuals the lengths of these stages are more or less definite, and death is seldom long deferred, but in the larger groups the final decline may be greatly prolonged. Why this should be is not at all clear, but one effect of it is that 'relicts' of some groups survive long after the apotheosis of their kind, and it is partly because of this that most floras and faunas comprise examples of so many different levels of evolutionary development. In some very remarkable instances of this persistence the group concerned appears to be almost everlasting.

But this waxing and waning in living things is by no means a process peculiar to them and there is good reason to believe that this kind of change is a very widespread phenomenon. Many of our basic astronomical concepts indicate that it is to be observed far beyond our own planet and, indeed, encourage us to believe that Earth itself is passing through a simple sequence. Indeed what we know of the cosmos as a whole gives ground for suggesting that such cycles of waxing and waning are a fundamental characteristic of it.

The considerations and arguments set out so far lead to one ineluctable final question. Has the initiation and subsequent course of evolution been fortuitous or has it followed some over-all master plan? Has it been the result of chance or the fulfilment of a deliberate design? If the latter where lies the authorship of the plan?

This is the great dilemma that evolution presents. On the one hand; given the existing and natural enough anthropomorphic view of the living world it is difficult to avoid the assumption that mankind, with its unique mental power, has been the supreme achievement of biological evolution and that all other of its expressions have been either its harbingers or its side products. As long as this attitude of mind prevails any belief that evolution has been fortuitous is surely calculated to diminish the stature of man in his own estimation, as well as in relation to other animals, by denying him that special quality of pre-eminence which has been, and still often is, regarded as his manifest destiny. It labels his emergence as no more than the outcome of chance, and as but one of the no doubt countless and unknowable 'might have beens', some of which would have been, perhaps, even more astonishing. On the other hand, to believe that evolution has followed a pre-determined plan directed towards a pre-destined end must involve a belief also in a designer of some kind, either in the more direct sense of a controlling intelligence or in the less direct sense that one step or event has led inevitably to the next, and finally to the state of awareness which puts all other nature at man's disposal.

How far either of these possible explanations of what appear to be the facts satisfies, must, in the last resort, be a matter of personal opinion based on rational and logical consideration of the information available, but it helps to remember that until man, by the growth of his intelligence, achieved some special power over his fellow animals and over the plants on which both he and they depended, he was, in strictly biological terms, of no more account than any other kind of living thing, and it is largely his own estimate, not in every way supported by fact, that has made him regard himself as the be-all and end-all of evolution. With this attitude of mind it is not unnatural for him to incline towards a planned explanation of evolution, with himself as its foregone culmination.

It helps also to recapitulate briefly some of the most salient of what we accept as the facts. The world on which we live has no substantial claim to uniqueness; there may be others in which similar conditions exist or have existed, but its outstanding features are plain to see. Nature, using that word to comprehend the world and 'all that therein is' is of two kinds. One, the mineral or inorganic, is vastly more extensive than the other and is devoid of any life. The other, organic, is far less pervasive but is *alive*. It lives because it is able to harness a certain amount of solar energy which would otherwise have become degraded and useless according to the Second Law of Thermodynamics and the principle of entropy. It continues to survive because it has succeeded in postponing the inevitable. This it does by means of peculiarly specialized processes involving complex chemical reactions unknown in the mineral part of nature.

As has been mentioned however, life is often regarded as unique, with the inference that it occurs only on this planet, and this has had much influence on ideas about its origin, because if it is indeed unique and nothing like it has ever occurred elsewhere in the universe then the likelihood that it has had an equally unparalleled kind of origin is much greater, and it is not too much to say that a belief in the uniqueness of life and of its leading expression, man, encourages the idea of some super-natural or mystical intervention, because it flatters our intelligence and sense of importance to suppose that we and our predecessors have been the special and peculiar subject of some matchless power.

In fact there is very little reason for supposing that life

has never existed anywhere else in the cosmos. The number of stellar systems is very great, and it is reasonable to think that each has something like its own system of planets. It seems likely, also, that each stellar system passes through the same kind of change as our own solar system. If this is so then it is likely that many, many planets or other celestial bodies have passed through the stage of change at which, on earth, life took shape.

Much difficulty here is due to a natural but exaggerated interest in our own few planets, an interest that has been much popularised in recent years. In fact the solar planets are very unlike one another in many respects and one of the easiest ways to account for this is to suppose that they are at different stages in their life-spans. Today only Earth among them presents the conditions under which life as generally understood can exist. Similarly on a vastly greater cosmic scale we may presume that if this is true of our own, solar, system it is equally true for other similar systems. The conclusion from this must be that, far from the conditions necessary for life having appeared only once and in one place they must have existed many times in many parts of space, and it is only a slight extension of this view to believe that every celestial body passes, as its energy declines, through a phase during which the conditions suitable for life are at their most favourable. If this is so then the thesis that life on earth has come about by the more special action of an external and super-natural agency takes on a rather different complexion. It may indeed by true but if it is then it is most likely to have been true also on many other occasions elsewhere, and the idea of uniqueness, which is one of its chief supports, cannot be sustained.

It can of course be argued that however many times life may have originated it has always been the result of the intervention of what it is perhaps easiest to call 'mystical' forces, but clearly the more often this has happened the less unique must the forces be reckoned, and the logical extension of this is that such agencies are not super-natural at all but a normal feature of the cosmos and of the life in it that they engender.

This conclusion is strongly supported by what is generally accepted as the way in which the evolution of mystical ideas gradually developed in the earliest stages of man's dawning intelligence. The first phases of this was his acquisition of the ability to associate cause and effect and to recognise past, present and future. As soon as he could do these two things he was faced with a very difficult problem. Not only did it become clear to him that there were eventualities inherent in his situation which were beyond his control, but he was able to see that these might produce crises with which he would be unable to cope unaided. Clearly if his anxieties were ever, in the long term, to be assuaged, he had to believe that there was some power outside and beyond himself to which he could appeal for help. This he originally did by supposing that all the expressions of his environment, some of which were beneficial to him while others were hostile, possessed personalities to which appeal could be made. This outlook narrowed itself, as time went on, into the concept of one or more gods, supreme because they possessed powers much greater than those of man himself, and from this idea the various formal dogmatic religions were gradually shaped. Hence man, whenever, since the earliest days of the growth of his consciousness he has been faced with apprehensions of possible future dangers, has tried to avert them by asking the help of whatever god or gods his fancy may dictate. This deeply seated instinct to appeal for help from outside has spread beyond the scope of mere physical dangers and has long been extended towards the solution of rather different problems which, to him, are incomprehensible and apparently beyond his control.

These, to him baffling, situations have been a recurrent feature in the gradual growth of knowledge, and on many occasions have led man to have recourse to his gods, with varying results. Sometimes the effect has been unfortunate and has led, not only to much religious persecution but also to retrograde steps in human development, but at other times the appeal to higher courts has, fortunately for the human race, been rendered unnecessary by some great, and often unexpected, leap forward in knowledge, so that what has for so long seemed baffling is seen to have a reasonable explanation after all, and the mysteries of yesterday become the foundation stones of tomorrow. One, and perhaps the most relevant, example of this is the potential power deriving from the fission and fusion of atoms which, until not much more than half a century ago was unrealised, but which, for better or for worse, has since transformed the world as a habitat for man, as well as much of his knowledge of it.

The situation today regarding the problem of the way in which life orginated on earth bears all the hall-marks of just such a bafflement. It cannot be explained fully in terms of existing knowledge but this is a measure not of its literal incomprehensibility but of the imperfection of our awareness, and there is every reason in precedent to believe that in due course this deficiency will be made good. No-one can accurately forecast the form in which this revelation will come but the indications certainly are that it will take the form of some new, perhaps completely new, concept relating to the theories and laws of energy.

Conclusions

This book is divided into four parts of which the first three are in one way the reverse of the fourth. The former constitute a brief survey of what is known, or believed to be true, about evolution in its more familiar meaning of the developmental history of the plant and animal kingdoms. It goes back to the earliest circumstantial evidences, and may be described as made up of fact seasoned with speculation. The last part is an attempt to probe still further into the past and is therefore largely theoretical, and it may be described as speculation seasoned with fact. Speculations, in their more dignified form of theories, have always been the first steps to better understanding, and to this the subject of biological evolution is no exception, and it now remains to try and blend these facts and theories into a single outline sketch of the way in which the world and its populations have gradually reached their present states. For this three cardinal points afford a reasonably reliable starting point.

The first is that the planet Earth of the solar system is restless, unstable and ever-changing. Opinions differ considerably about the nature and sequence of these changes but as far as any discussion of biological evolution is concerned it may be assumed that one of the most important of them has been, for an immense length of time, a slow and more or less continuous loss of heat, and the world may justifiably be regarded in this context as having gradually cooled from heat so high as to be difficult to appreciate to its present temperature values.

A second cardinal point is that the earth possesses

water, that most peculiar substance which is liquid only within a very narrow temperature range close to the absolute zero of the thermic scale.

These two cardinal points control all other influences, but as far as biology is concerned there is a third point almost equally significant, which is that mundane life (and it is necessary to specify this because it is not known what other forms of life there may be elsewhere) expresses itself through the medium of dilute aqueous solutions. There is no active life divorced from these.

These three points in combination lead to the first incontestible conclusion about organic evolution on earth which is that it cannot have begun until what are, by cosmic measure, recent times. The corollary to this is that the world and, as far as is known, the universe, was lifeless until almost the last short chapter of its long, long history.

From this broad outline can be drawn the further conclusion that life, as it is familiar today, could not have originated on earth until liquid water was present, that is to say until any water vapour which there may have been in the surface layers of the earth's rocks or in the gaseous zone above them had become condensed into its liquid form. This is tantamount to saying that life could not have come into being until the temperature of the earth's surface had fallen to the point at which water vapour becomes liquid. The question is how, more precisely, life originated in these circumstances?

The waters of the oceans, formed at the appropriate temperature by the condensation of the water vapour in the atmosphere would, at first, have been free from mineral salts but would have contained ingredients of such atmospheric gases as may have been soluble in them, the most important of these being carbon-dioxide which, with water, forms the weak and unstable carbonic acid. As soon, however, as there was liquid on the earth's surface, and precipitation from the atmosphere, the erosive effects of the water and any solutes that it may have contained would have begun to add other constituents to the ocean waters, not only from the land surfaces by subaerial denudation but also from the bed-rocks of the seas, then unprotected by any superficial deposits. Thus, from the beginning, the 'impurity' of the sea water increased. Just how these events shaped themselves is not clear but it is believed that as time passed the number, and up to a certain point the concentration, of these impurities increased.

It is from this belief that there has developed what is now the most generally held view about the way that life originated on Earth, namely that which it is convenient to call the 'soup theory'. This supposes that the water of the oceans gradually became a more and more complex aqueous chemical solution and that, in this process, ever larger and more intricate carbon compounds were formed, and more intense reactions took place between them, until at long last some of the molecules became combined and organized into entities which exhibited some at least of the reactions characteristic of active life.

This theory is attractive not only because it is simple and because there is no obvious alternative to it but also because it seems to evade the problem of whether or not life began through the agency of some super-natural power, but this is a false conclusion, and the two are not mutually exclusive because the vehicle of change as outlined in the theory could have been the means by which an external power manifested itself.

Moreover, the theory is imprecise in an important respect. Expressed in its simplest and most common form it carries with it a strong implication that life originated on earth at one particular time only, though not necessarily at only one particular place, but this is a narrow concept which derives its emphasis from the influences of traditional creation stories, and other possibilities must not be overlooked. Some features of the living world are most easily explained on the supposition that life has originated more than once at comparatively long intervals of time, but even more important than this there is no definite evidence against the view that life has never, since its first inception, ceased to come into existence. Nevertheless the 'soup theory' in some form or another is most generally held to reflect the sequence of events which, beginning with the formation of the oceans, eventually led to the appearance of living things.

Unfortunately the simplicity of the theory is deceptive. It is easy enough to visualize the very early stages in the accumulation of the products of erosion but this alone does not take the matter very far and it is the blank wall into which speculation runs at the end of this stage which is the most puzzling of all evolutionary problems and the one in which ignorance is still most conspicuous.

The difficulty may be put in this way. The 'soup theory' almost inevitably gives the impression that life finally emerged as the end product of what has, in quite another context, been called the 'inevitability of gradualness', and that the passage from the mineral to the organic was a smooth progression, but this view tends to ignore or minimize the difference between the two states, or to avoid the issue by using such phrases as 'the vital spark'. In fact life is the performance of certain functions, and the ability to *do* as well as to *be*, and is fundamentally different from non-life because it involves the manipulation of energy. This is because the world is, through the dissipation of the various forms of energy inherent in it or associated with it, continually moving towards entropy and whether the eventuality of total entropy is a reality or not it is accepted that the world is, to use a simple phrase, slowly 'running down'.

This is one of the crucial points in the evolutionary story because, in order to carry on their vital processes, that is to say to *do* things, living bodies must be able, directly or indirectly, to acquire energy from sources external to themselves and to apply it to their own needs. They must be able to tap the energy sources of their environment and to harness them for their own purposes.

In the biological world of today this constant rewinding of what may be called the energy clockwork is done by green plants which possess, in their chlorophyll, certain chemical catalysts that enable photosynthesis to go on. This sounds simple enough but photosynthesis is a highly sophisticated process and, and although a good deal is known about it, there is still ignorance of just what the catalysts in chlorophyll are and how they work.

This reference to chlorophyll is intended to emphasize the complexity of the chemical substances involved and the essential part that catalysts play in them, because photosynthesis is not the only process controlled by such agents. Some of these other catalysts are better known, and some of them, such as the oxidases that control respiration, are enzymes. The point is that all these substances are highly complex chemically and the difficulty is to understand how the earliest of them can have emerged in the course of events suggested in the 'soup theory'. Today these catalysts are found only in living bodies. They are an expression of life and in the absence of it they do not exist. The puzzle is that while it is difficult to suppose life without them it is even more difficult to imagine them in absence of life.

It is not known just how catalysts of this sort made their first appearance, and ignorance about this is one of the more noticeable of the many blank pages in the story of evolution. Indeed the gap is even larger than might at first be thought because the 'soup theory' is highly speculative and there is little evidence to support it. This is not to say that the course of events suggested in it did not occur but it does mean that the blank pages are even more numerous than might be supposed, for our lack of knowledge extends all the way from the time before any life existed to that at which living things in fairly advanced form were first entombed as fossils. This 'Great Gap' represents an immense space of time and virtually nothing is known about it. It provides time for the most surprising and unforeseen things to have happened and for the possible intervention of forces beyond the normal range of comprehension and even, perhaps unknowable by the human mind. It is certainly hard, in thinking about these early stages, to resist the impression that lack of information may be due not so much to lack of factual knowledge in one or more of the traditional sciences as to lack of experience in other fields of enquiry. This is at least a worthwhile thought with which to leave this puzzling period of the earth's history.

It is something of a relief to pass from these speculations to the firmer ground of the fossil record and here there are two points in particular to note. The first is the likelihood, based on a general appreciation of what is known about the history of the planet Earth, that the time which elapsed between the first appearance of life on it and that at which the first fossils were preserved, was longer, perhaps much longer, than the period that has elapsed since. If this is so then the whole of the evolutionary process as expressed in the fossil record and in the characteristics of plants and animals living today must, in the dimensions of cosmic time, have taken place with great rapidity. Coupled with this are the indications given by the fossil record that the speed of evolution, as measured in the frequency with which new forms have arisen, has long been increasing.

The second matter is fundamental to the whole idea of organic evolution and may even be the master-key to it. The world is changing, and has long been changing, because it is a gradually cooling member of the firmament, and this change is its over-riding characteristic. In the ages before life appeared, as well as since, this change has been accompanied by a relatively simple and gradual degradation of the world's energy towards the condition of entropy. Many detailed changes of many kinds occurred in the ages before life appeared but these were essentially kaleidoscopic, consisting of the rearrangement and redistribution of matter and energy as then expressed. With the coming of life this state of affairs became fundamentally altered by the ability of plants to slow down this movement towards entropy in varying measure by harnessing some of the energy which would otherwise continue towards degradation for the particular purpose of multiplying, by biological reproduction, the very agencies able to delay it. In this way matter was not merely redistributed in space but recast into kinds hitherto unrepresented. While the immemorial kind of change with time continued in the mineral world (though not entirely unaffected by the presence of life), in the living world it took on a completely new complexion. There must even be taken into account the possibility that the whole existing energy system of the world was altered and that such basic principles as the conservation of energy were disrupted.

The mineral world in pre-biological times was probably never, at least after its original solidification, completely homogeneous. Some areas must always have presented more opportunities for the potential development of life than others. To take but two examples, the chemical constitution of the surface rocks must have differed from place to place, and so must the depths of the ocean waters, but until some form of autotrophic organisms developed these differences were of little account. But when, by virtue of the catalysts they possessed, these beings were able to harness energy for the purposes of their own sustenance, the local differences became of great importance, and there was thus introduced into the world economy a quantitative factor as a result of which not all the earliest autotrophic organisms were equally well situated. Because of this the first autotrophs, even if they were planktonic, found themselves, not in a consistent environment, but in one which varied in potentiality from place to place and from time to time. This was, of necessity, reflected in these autotrophs themselves, not only because they were themselves measures of their environments but also, because they had acquired the power of reproduction, in their progeny.

Reproduction is the whole foundation of biological evolution, and the first point about it to be emphasized is that the general elaboration of living bodies, presumably on the basis of protoplasm or its precursor, itself involves reproduction, which is the ability to give rise to new, separate, and usually more numerous living bodies, and the question of how this ability can have been acquired is a basic biological problem.

The mere production of progeny is sufficient cause for wonder but looked at a little more closely it is seen to consist of two remarkable features, namely a multiplication factor and the organization of biological life into life-spans. The former is the easier to rationalize because something of the sort is clearly necessary if the biological lineage is to survive the destructive elements in its environments, and this multiplication factor is often of remarkable proportions. The latter is more mysterious. Why should life be determinate, and why are plants and animals born only to die? The answer is clearly connected with the multiplication factor because, given this, some determination of life is clearly necessary to prevent a degree of over-population which would soon lead to disorder. If progeny are to succeed (in both senses of that word) there must be a constant 'dying-off at the top' to allow for it, and to prevent the undue accumulation of living generations. This argument can be reversed. If individual organisms did not possess the power of reproduction death would ultimately prove racially fatal. It is perhaps in the puzzles of reproduction, and in the remarkable niceness with which the whole organisation of life and death operates, that there resides the kind of problem to which the ordinary rigid laws of chemistry and physics do not provide the answer, and which therefore provide the strongest reasons for the opinion that something more, which ensures that life shall and should endure, is involved in the affairs of nature.

But to return to the subject of the earth itself which provides the environment for all its living inhabitants. The world, because it is a celestial body generally described as having long been cooling, is itself undergoing a constant slow evolutionary change. This planetary change appears, in certain respects, to be a slow decline in certain values, while the biological evolution that attracts so much more attention appears to be a process of aggrandisement, but the latter does not counterbalance, or even seriously hamper, the former, and despite the activities of the living things on it the earth is moving towards energy exhaustion. What may be described crudely as the cooling of the earth has, to the best of our knowledge, always been, on balance, a one-way process which, though it may not have been an even one, has never been subject to any but possibly short-term reversals, and hence all the more superficial changes and effects accompanying it are to be regarded as part and parcel of it.

The initiation and elaboration of mundane life can only be regarded as such a superficial effect and hence biological evolution must in the first instance be a slow change and in the second a change in an irreversible direction. In short, given a world that is constantly changing in one fundamental respect and direction some form of what we call evolution is inevitable because the inhabitants of the world are bound to be influenced by the effects of this change and the principles controlling it.

This may explain why there has been evolution but it does not explain why its biological expression has been what all the available evidence reveals. The answer lies in the word 'variation'. The great difference between the members of the living world and the materials of the mineral world is that the former depend for their existence on their ability to derive sustenance from the contents of the latter and the chemical mechanism that enables them to do this is sophisticated, involving the evolutionary development of special catalysts. We do not know just how this catalytic metabolism first began but living bodies must have been evolving for a long time before they were able to accommodate even the simplest of such systems, and the bodies which first exercised it must have been relatively complex. More than this they must have possessed an element of variability because their environments were not all alike. Some would have provided the necessary materials in abundance, others but sparingly, and these differences must have influenced not only the quantity but also the quality of the catalytic synthesis. This situation resulted in the phenomenon, not present in the same sense in the mineral world, of variability, resulting in that inequality which distinguishes one individual from another.

As these catalytic systems and the organic bodies that operated them became more and more intricate with the passage of time, so the likelihood that any two living things are ever the exact duplicates of one another correspondingly decreased and it is now regarded as a reasonable assumption, supported by all common observation and experience, that no two individual plant or animals are ever quite alike, whatever their relationship in time or lineage may be, and hence that every newly formed individual living thing is a novelty. *This variation is an essential feature of the organic world and, coupled with the principle of irreversibility in a changing world is, alone, enough to produce that slow and gradual change in living things which is called biological evolution.* This being understood, the unlikeness between individuals of the same generation and between parents and offspring becomes the most fundamental of all considerations relating to evolution, and the question of how it may have come about becomes of great significance.

Basically the reason is that the development of living things introduced a new variable into the existing world system, which, on that account, became no longer exclusively subject to the laws of chemistry and physics as before. This variable is plant and animals metabolism which, because of the heterogeneity and complexity of mudane environments, together with the passage of time, results in a fundamental variability which makes the likelihood of total similarity between individuals infinitely remote.

At this point let us recapitulate. If the picture of Earth as a slowly changing planet is correct its story can be told in two chapters very different in length. Throughout the earlier and longer chapter, which lasted until the lowering of temperature had reached values which cannot have been very different from those of today, the earth was lifeless. Then, in circumstances at which we can only guess, there appeared the first entities exhibiting vitality. It is supposed that the earliest of these were of the simplest design and life-pattern but in the fullness of time they became more elaborate and multifarious. Changes in the life-pattern also took place, consisting of a gradual crystal-lization into a standard procedure from which, as far as is known, there has never been any appreciable subsequent departure. This procedure is the generational organization of life by which, at intervals, parents hand the torch to their progeny before themselves leaving the stage. The life of every individual is determinate, though it may vary much in duration, and birth is but the precursor of death. That this inexorable death has not so far been ubiquitous and simultaneous enough to entail the extinction of life is due partly to differences in life-spans but mostly to the safety mechanism provided by reproduction. This has two rather different functions. More important it gives continuity to life by producing new bodies before time and circumstance prove fatal to those already in existence; but it is also a form of multiplication which protects the lineage by providing insurance against the hazards inseparable from such a finely adjusted sequence of events as life.

That the period of individual life is determinate is one of the most singular aspects of biology. In some ways it is reminiscent of machines which operate as long as their supplies of fuel are not interrupted and which themselves wear out only after a comparatively long period of activation, but this comparison is incomplete. Particularly, living organisms have an 'inner' cycle of life unrepresented in machines. Except for some of the simplest organisms the first phase of life is one of growth, at the end of which the condition of maturity is reached. Then there comes a phase, usually longer, when reproduction is the primary activity; and at last there comes ageing, viability gradually dwindling to the nothingness that is death. Reproduction may continue little diminished until the end, and this seems to be the norm, but reproductive power may decline much earlier. Another peculiar point is that all living things begin as single cells and, in their early stages, grow by accumulating body material, but after a time this dimensional increase ceases and the activity of the organism becomes more and more centred on reproduction. This very pronounced cycle strongly suggests that, in the first stage, 'vigour' (it is better to avoid the word 'energy' here) is accumulated beyond the immediate needs of the individual; that in the second stage a considerable part of this vigour is used in reproduction; and that in the third, which is not always distinguishable from the second, such vigour as remains finally becomes dissipated and death ensues, the materials comprising the body decaying and eventually returning to the environment as 'ashes to ashes; dust to dust'. The belief that something, here called vigour, is acquired in early life and frugally drawn upon later is strong but difficult to explain in terms of the conventional laws of physical energy.

Nor is this all. In the simplest organisms, such as the bacteria, reproduction consists simply of the division of a single parent cell into two separating parts and the restoration of each part to the original parental dimensions (this is 'replication without diminution'), but the gradual elaboration of plant and animal structure, which is the most obvious expression of evolution, has been accompanied by increase in the complexity of the reproductive process. In briefest outline the simplest reproduction is asexual, being the formation of propagules which are the products of the division of but a single living cell, but even in many simple plants and animals this is superseded by sexual reproduction in which each propagule begins as the consequence of the fusion of two parental nuclei. Again, among comparatively simple organisms these two nuclei (gametes) are alike but very early in the evolutionary pageant these become differentiated, until, quite early on, the full system of sexual reproduction, involving passive egg cells and active sperm cells, became established. The subsequent variations in, and refinements of, this system, form a large part of the evolutionary story.

Difficult as it is to believe that there has been a fundamental reason for all this it is even more difficult to believe that there has been no reason for it at all, and this raises a particularly puzzling point. Does this remarkable system of sexual reproduction exist because there is, in the overall plan of nature, an inescapable need for it, or is it gratuitous in the sense that it is an unnecessary complication of a function which could easily be performed with a simpler machinery?

In trying to answer this question, which is only a more particular form of the query as to how life began, we can at least be sure that if the complicated processes of sexual reproduction had evolved in some different way the state of the living world would not be that of today. Can we go further than this? Can we believe that the system has been *intended* to produce its observed effects? Can the word *intention*, in any one of its several rather unsatisfactory definitions, provide some kind of option additional to pure chance on the one hand and creation on the other?

It was said earlier in these 'conclusions' that the 'soup theory' is attractive because it seems, at first hearing, to evade the issue which has complicated the study of evolution ever since that began in earnest, of whether the origin of life on earth was due simply to the operation of scientific laws or whether it was brought about by the intervention of some external and 'super-natural' agency. Some prefer to believe the former, some the latter: others, no doubt, satisfy themselves with a combination of the two, or prefer to express no opinion.

This question has been discussed time and again and from many points of view, but with little positive result, and a book on biological evolution is not the best place in which to argue the matter yet again at any great length.

Nevertheless it would be wrong to make no reference at all to it, because despite the great advances in knowledge in nearly all branches of science it still remains the principal biological uncertainty, and the most helpful way of doing this is to remind the reader of the outline, given at the beginning of Part II of this book, of the way on which ideas about the supernatural first, and inevitably, emerged in the very earliest stages of the growth of man's intelligence.

Any account of evolution is bound to be couched so much in terms of the past, and especially the more distant past, that it is easy to forget that it is a continuing process still in action today and likely to pass into an as yet undisclosed future. Some hint of what this future may be emerges from a recapitulation of the last few pages. The whole long process of organic evolution has been made up of three phases, each shorter than the one before. During the first, which lasted for an immense period, the process was, in comparative terms, slow, and confined to an aquatic medium. Then, at what is commonly reckoned to be some few hundred million years ago, plants and animals of the water gradually invaded and occupied the land in what is called 'the subaerial transmigration'. By this a threshold of immense significance was crossed and it was followed by a great surge of evolution in the course of which the unfamiliar subaerial environment was exploited, a process that produced not only great numbers of new morphological designs but life in many new expressions. When the biota of the sea is compared with that of the land it is not difficult to see that the development of the latter has been marked, not only by the multiplication and elaboration of form which came about in the course of filling the almost infinitely varying niches in what was, at first, a biological vacuum, but also by the intensification of instinctive behaviour called for by the fundamental change of habitat and the necessity of responding to its many novel aspects.

The second phase may be considered as continuing until there arose living things possessing, in addition to

their inherent instincts, the first glimmerings of intelligence, no doubt first manifested in an ability to relate cause
and effect and in the recognition of past, present and future
time. This initiation of intelligence is the second great
threshold that evolution has crossed, and following it there
gradually appeared on the scene living animals capable of
ordering, or disordering, their lives by the exercise of
deliberate conscious action. The primary evolution of the
first and second phases is still continuing but there has now
become superimposed on it a third phase.

This is the time, measurable perhaps in only hundreds
of thousands of years instead of many millions, which has
marked the advent of man. With his coming the whole
nature and future of the world has been changed, because
this unique product of the evolutionary process on earth
has achieved the power, not only to choose but to
implement the choice, so that he is able to modify the
conditions of his environment to favour his own ends. One
aspect of this is of supreme importance. The evolution of
humanity has produced with it a new kind of change which
has been made possible by man's realization that he has the
power to take actions, and to make objects and conditions
which, he calculates, will improve his lot.

The consequence of this is that although evolution of
the more traditional sort is still going on in the plant and
animal worlds, the most marked evidence of change with
time in the human sphere is the evolution of man's ideas
and artifacts – the products of his intelligence and of his
tools. The extraordinary feature here is the astonishing,
and apparently increasing, speed at which the process is
moving, so that the world as a biological habitat is plainly
changing before our very eyes. Man has indeed become the
'lord of creation', with all the powers of life and death
appropriate to such a potentate, and the point is now being
reached at which doubt about the future, supported by not
a little evidence, is beginning to arise. Are man's powers
leading him togards a condition of ordered progress or
towards a state of chaos? If the latter will it be possible for

him, by the proper use of his immense intellectual and material resources, to correct the movement before it becomes completely beyond his control?

The foregoing pages are a condensation of the four preceding parts of this book and of the conclusions that may be drawn from them, but it is itself of considerable length; needs to be read with close attention, and covers a variety of constituent topics. It can, however, be further compressed, in most respects, into the following statement.

The gradual change in plant and animal form and function that is called biological evolution results from a combination of two circumstances. One is that the world has long been changing, principally in the direction of temperature alteration, and that this process has continued since it became the environment of living plants and animals. The other is that no two individual living things are ever precisely alike, nor can they be nor ever have been, because the passage of time is one of the determinants of change. The result is that every new individual is a *novelty*, different from any one that has gone before or which may come after. These two circumstances are sufficient in themselves to account for biological evolution but they do not explain why that process moves, as it so patently appears to move, in the direction of continuing, and apparently increasing, diversity and elaboration. Clearly there is operating somewhere in the system a principle of accumulation which determines that the effect of change once brought about and of time once elapsed, are never thereafter wholly lost. Change may take new directions but, since no individual is exactly like any one of its forebears or its posterity, and since the passage of time, with its attendant eventualities cannot be nullified, evolution can never be truly reversed.

Evolution thus comprises two components, time and change, of which the former may be regarded as a constant and the latter as a variable. If change is slow evolution will be the less apparent; if it is rapid more evolution will be expressed. It is this variation in the speed of change which results in the multifariousness of living nature and it explains why every biota of any size contains examples of many levels of evolutionary development. Living plants and animals therefore express in their characteristics not only the degree of change that has taken place in their lineage but the length of time that this has involved, though the two need not necessarily be directly related.

Biological evolution can therefore be regarded as the structural and functional embodiment of accumulated association with, and experience of, a changing environment, which the longer it continues the more comprehensive and all-embracing it becomes. Just as in the realm of human affairs no living person can at any time become *less* experienced than he was before, because experience is the direct and cumulative effect of a passage of time that cannot be reversed, so in the rest of the plant and animal worlds no new individual can be without the inherited evolutionary past which is the expression of its particular genealogy but, unavoidably, must carry the account forward. It is to these influences that there must be ascribed those observable effects to which the name of biological evolution has been given.

Index

3/16